HIGH ENERGY

HIGH ENERGY

How to Overcome Fatigue
and Maintain Your
Peak Vitality

ROB KRAKOVITZ, M.D.

JEREMY P. TARCHER, INC. Los Angeles
Distributed by St. Martin's Press   New York

While the information in this book is based on sound medical and clinical research and experience, it is not meant to substitute for a consultation with a medical doctor or other health professional. If you have any known medical problems, speak with your doctor before treating yourself through nutrition or with any other methods.

LIBRARY OF CONGRESS CATALOGING IN PUBLICATION DATA
Krakovitz, Rob, 1947-
  High energy.

  Bibliography: p.
  Includes index.
  1. Health. 2. Vitality. 3. Fatigue—Prevention.
I. Title.
RA776.5.K67 1986   613   85-27693
ISBN 0-87477-374-1

Jeremy P. Tarcher, Inc.
9110 Sunset Blvd.
Los Angeles, CA 90069

Design by Cynthia Eyring
Manufactured in the United States of America
S 10 9 8 7 6 5 4 3 2 1
First Edition

For all the tired people who never knew why. . . .

For Sylvia,
          who always loved and believed in me.

For Bernie,
          wherever you are.

ACKNOWLEDGMENTS

I wish to express my appreciation to:

My patients, for providing me the basis and inspiration for writing this book;

My friends, for their love, support, encouragement, and enthusiasm;

Mitchell Chipin and my office staff, for their help and support in running my practice;

Henry Bieler, M.D., Alan Nittler, M.D., and Joseph Walters, M.D., for their wisdom, courage, and vision;

Lazaris and *A Course in Miracles,* for inspiring and motivating my personal growth and for the bases for many of the metaphysical principles and techniques passed on in this book;

    Connie Clausen, for her support and encouragement;

    Janice Gallagher and Jeremy Tarcher, for seeing, knowing, believing, and working with me to create this book;

    Dominick Bosco, for his friendship, loyalty, focus, determination, and dedication in helping me to prepare the manuscript and make this book a reality.

CONTENTS

INTRODUCTION

Do You Want More Energy?

Do you have all the energy you want? Do you get tired at certain times of the day—for no obvious reason?

Have you gone to your doctor with this rather vague complaint, only to be told "there's nothing wrong with you," "it's all in your head," and "you should learn to live with it"?

No matter how many doctors have told you this, I know that your fatigue is real and that it's curable. You don't have to ignore it, learn to live with it, or wonder any longer why you're tired. Before you read another 25 pages in this book, you are going to know what's causing your fatigue and be on your way to knowing what you can do to get more energy and keep it.

Every day, in thousands of doctors' offices, millions of

men and women complain that they don't have enough energy. Fatigue is the single most common complaint most doctors hear. But seldom does anything abnormal show up on the laboratory tests of patients who complain of fatigue. Conventional diagnostic tests are not designed to pick up the subtle, but insidiously powerful, physiological imbalances that cause most cases of fatigue. Fatigue exists in the gray area between "feeling wonderful" and "defined illness," and most doctors simply aren't trained to deal with symptoms and influences that go beyond the textbook.

You don't need a doctor or a book to tell you you're tired. You know that already. What you don't know is *why* you're tired. This book will tell you why you're tired and what to do about it.

My high-energy program will deliver. I know, because I've used it on hundreds of patients in my office over the years as I've fine-tuned it to its current level of effectiveness. It's worked for my patients. It will work for you.

HIGH ENERGY

1

Your High-Energy Future Begins Right Now

"Give a man a fish and you feed him for a day. Teach a man to fish and you feed him for the rest of his life."

PROVERB

Randy was known around the office as a dynamo. Only 10 years out of business school, she was already vice president of a growing publishing company. Everyone agreed that when she walked through a room, the air seemed electrically charged.

What they didn't know was that every evening, once the door of her condominium was safely closed behind her, Randy would collapse on the sofa. Hours later she would get up and drag herself to bed, where she would sleep restlessly. In the morning, she would get up in a fog and only a hot shower could even begin to wake her. Three cups of strong coffee would boost her out the door. By the time she got to the office, her secretary would have a fresh pot perking. Every time Randy felt the fog closing in around her, every time she felt as though she might not make it to the end of the day and her sofa before she collapsed, she would drink a cup of coffee. In the course of the day, Randy would drain the four-cup machine three times.

When Randy came to me, she had already been to four physicians and was desperately seeking a stronger stimulant to give her the lift she needed.

Fortunately, Randy didn't need any drugs to get high energy. By following my program, she learned that the coffee she believed was giving her energy was actually stealing it. She was able to free herself from what amounted to an addiction to coffee and other stimulants and rebuild her energy by altering her diet and lifestyle. Randy now gets more work done than ever before, and she has energy to spare for exercise and an active social life.

Bill, a very successful man by anyone's standards, was entering his 40s with the prospect of continued creative growth and success in his production company, which had progressed from making TV commercials to feature films. And yet, when Bill came to me, he was quite depressed and confided that he didn't know what was happening to him. "I've always been completely healthy. I've always had energy to spare, but lately, these last few years, I've noticed that I can't stay up as late and get any work done. I get lazy in the middle of the day, too. I eat only natural foods, no sugar, coffee, or alcohol. My medicine cabinet doesn't even have any aspirin!"

Bill had been to three other physicians. They had all told him he was "just getting older."

But Bill is no longer depressed. He's no longer in the dark about what goes on in his body, and he tells me he has more energy than he had when he was in college. How did he bring about this change? Some of Bill's favorite foods—although perfectly natural—were doing him in. Not only did Bill need to sharply de-emphasize these foods in his diet, but he also needed to stoke up his energy fires with some regular exercise. Now, instead of slowing down in the middle of the day, Bill can be found rolling along the bike paths on his daily five-mile jaunt.

Sally was approaching her 35th birthday with anxiety. Her son and daughter were teenagers now. Their grades were excellent and they helped around the house. Sally's husband's career had reached a successful plateau and he had taken up the fine art of cooking. Sally took up jogging and started taking nutritional supplements. She had long ago thrown all junk food out of the house.

When she came to me, Sally told me she had tried everything—exercise, natural foods, and supplements. Still, she did not know why she felt her body was about to "crumble in a mushy heap." There were times, she said, when she felt afraid and times when she felt angry. The only thing she had found that helped at all was to take a nap. Lately she had found herself napping three or four hours a day.

Sally's problem was not entirely physical. She did have some food sensitivities, which were easily handled through my detoxification and desensitization process. But the primary drain on her energy was emotional and psychological. Sally was suffering the early stages of the empty-nest syndrome. She needed to focus some of her truly abundant inner energy on her own creativity. Sally now runs her own athletic clothing shop. She no longer needs those naps. To say she has high energy would be putting it mildly.

All of these people were suffering from fatigue. Most of the time, fatigue means simply not having enough energy to do all you want to do. But sometimes, fatigue manifests itself in other ways. You might feel only a vague sense of not having enough "power" inside you to get a full day's work done or to withstand all the stresses and strains in your life. Or you might have a sense of expending great amounts of energy and getting nowhere. Paradoxically, fatigue often masquerades as high energy. You might seem energetic but feel depressed, anxious, or bored. Perhaps you feel as though you have the strength of a locomotive—but your brain feels too fogged to concentrate that energy. Or perhaps you depend on stimulants to keep you going.

Your sex drive might be suffering. When you drop out of the plane of high energy, one of the first things you usually miss is libido. So if your primary problem is lack of sex drive, it's possible that your fatigue is just beginning. Whenever a patient comes to me with this problem, once I've ruled out other physiological causes, the energy-building steps in this book become a sexual energy restoration program.

Maybe none of the above-mentioned problems are bothering you. You may just have a feeling that you would like more energy than you now have. This, too, is a valid reason to consider the high-energy program. In my years of medical practice, I've learned to trust a patient's feelings and self-concept. Even if you don't suffer from tiredness, brain fog, depression, difficulty concentrating, weakness, or lack of libido, but you do have a sense that you would like to have more energy and more confidence in your ability to maintain energy in order to get more accomplished in your life, this book will help you achieve that goal.

THAT HIGH-ENERGY FEELING

How will you feel when you have high energy? High energy strengthens you inside and out, from head to toe. Your mind will be clear. You will have confidence that you can take on the challenges in your life and win. You will have the ability to relax and know that the energy to work, play, and think will always be there for you. You will feel comfortable with your body and its abilities. You will have a sense of power about your life and know that you are in control of the resources within you and able to make positive use of the resources outside you.

You will notice that I use the words "feel" and "sense" a lot. This is because high energy arises out of your mind as well as out of your muscles, bones, and organs. The relation ship between body and mind is a busy, two-way thoroughfare. What happens in your body affects what goes on in your mind. And what you think and feel affects what goes on in your cells, organs, muscles, and blood vessels.

You may have taken all of the physiological steps to increase your energy and still need to go beyond the merely physical to achieve high energy. You may have the best diet, the most complete nutritional supplement program, the most exciting and fulfilling exercise program—and yet the stress in your life may be draining all the energy your body can produce, leaving you tired. If that is the case, the energy your body alone can produce will never be enough. Successfully dealing with stress and the other challenges and opportunities life provides is not only a matter of nurturing a body that produces energy. It's also a matter of directing that energy into useful and effective areas. And the responsibility for that direction belongs to the mind, the emotions, and the spirit.

Energy is our birthright. Unless there are special genetic or congenital problems (which are extremely rare), we are born with the potential to have all the energy we'll ever need. So when you reach for that energy and it isn't there, it's a definite sign that something is wrong. Something in your body and mind is out of balance. Most people—including physicians—don't think of fatigue as an illness, or even as a precursor of illness. But is illness an abyss into which we suddenly stumble, or is it a hole we dig beneath and around ourselves, deeper and deeper?

In my medical practice I see a tremendous number of people who complain of being fatigued or of not having as much energy as they want. Yet these people, for the most part, are not definitively ill. They don't have degenerative disease such as diabetes, heart disease, arthritis, or cancer. They are in a gray area where they are not truly healthy yet not acutely ill.

Obviously, people don't generally feel wonderful one day and then suddenly have a serious disease the next. You don't have all the energy you need on Monday and suddenly find yourself crawling around the floor on Tuesday. Life simply doesn't work that way.

The way the body moves between health and disease can best be represented by a spiral. At the top of the spiral, you're healthy and you have all the energy you want and need. At the bottom of the spiral, you are in the acute and chronic stages of degenerative disease. The middle of the spiral represents the transitional area, where your body may not exhibit, and medical testing may not substantiate, any of the overt diagnostic symptoms of degenerative disease, yet you don't have as much energy as you would like. As you head down the spiral, your fatigue will increase, and the joy and happiness in your life will decrease.

Why is fatigue happening to you? The chapters ahead will answer that question more specifically. In general, as we get older, we need to take better and better care of ourselves if we want to maintain steady health or move up on the spiral. It's a fact of life that when we're young, the sheer power of youth enables most of us to get away with taking sloppy care of ourselves. But as we get older, the backup systems that kept us going start to slip away. We don't necessarily get weaker, as such, but we lose the wide margin of error that allowed us to remain strong even when we were doing noth-ing special to support and nourish that strength.

I want to make it clear right from the beginning that

fatigue is much more than a nutritional deficiency or a digestive problem. Energy and high performance cannot be restored and maintained merely by eating a diet high in complex carbohydrates and taking some nutritional supplements. This approach neglects the central problem. At the core of our strength, at the very center of our high-energy-producing mechanisms, are the adrenal glands. Hormones secreted by the adrenal glands and the thyroid gland stimulate and regulate just about every physiological mechanism that produces energy. As long as thyroid function is normal, the health of the adrenal glands, more than any other gland or organ, determines how much true high energy we can have (see Chapter 3 for a discussion of thyroid function). Stress and poor care weaken the adrenals and diminish hormone output. So as our bad habits begin to catch up with us, the adrenals try to bear the burden. Unless the adrenals are properly maintained, they will continue to weaken—and so will our life force.

Far too many doctors misdiagnose weakened adrenals as aging. Age in and of itself implies no obligation to fatigue or disease—but the older you are and the worse care you've taken of yourself, the longer this program may take to restore your energy. Your energy *can* be restored. To get it back and keep it, you must replace the sheer power of youth, which has faded, with the sheer power of your determination and good care of yourself.

Now is the time to do something about your fatigue. If you wait until you slip further into the acute or chronic stages of an illness, where other symptoms and fatigue are more intense, it will be an even longer road back up the spiral to the high end, where you have all the energy you want.

GOING THE DISTANCE

The goal of this book is to show you how to take care of yourself physically, intellectually, and emotionally in order to spiral back up to high energy. You'll learn how to acquire the confidence you need to make a difference in your level of vitality, to get back the vigor that was your birthright—and keep it. You'll learn to pay attention to the signs your body gives you, to catch yourself so you never move down the spiral too far, and to fine-tune your energy system to keep yourself at the high end.

Whatever energy level you're on now, my program will give you more energy. Wherever you are on the spiral, you will improve—whether you're so tired that you've been dragging yourself from doctor to doctor, or whether you're an Olympic-class athlete who wants to get that extra burst of energy for the competitive edge.

This book is the best program that I know of for regaining and building high energy. The object is not for you to become a fanatic, but to work all the positive steps into your life as best you can. Some of the steps may appear difficult. It would be wonderful if there were a magic pill that would instantly restore your energy. There is not. You did not lose your energy instantly and you're not going to get it back instantly. I will say, however, that many of my patients do use the word "magic" when describing how effective this program is in restoring their lost energy.

I do not expect everyone to instantly embrace or perform the entire program. However, I must make it clear that you may not achieve your highest energy potential unless you eventually take all the steps. As a medical doctor, it is my ethical duty to provide the best care I can. In this book it is my duty to provide the best information I can. I cannot withhold from you what I know will help restore your energy. It would be unrealistic of me to suggest that you can achieve your full level of health and energy without doing the entire program.

However, this is not to say that you cannot take the program one step at a time. Please understand that you may need to take more than one or two steps in order to see any results. I am certain that once you do begin to see and feel your energy build, you will be motivated to press on.

HIGH-ENERGY PROGRAM NOTES

Each chapter describes one complete aspect of the high-energy program. The chapters follow the same course I generally recommend for my patients. The first order of business is to find out what's causing your fatigue. The second chapter will do that by giving you a test that you will score yourself. Next, it's important to identify any problems that *must* be treated by a medical doctor. Chapter 3 will do that. If the wrong foods are draining your energy, either through a food sensitivity or low blood sugar activity, Chapter 4 will explain how this is happening. Chapter 5 will tell you how to rid your body of fatigue-causing toxins as you learn how to avoid foods that steal your energy. Chapter 6 will tell you how to rebuild your diet and your energy supply with the right foods. Chapter 7 will tell you how to supplement that diet to ensure high energy. Chapter 8 will explain the crucial role of exercise in building energy and tell you how to design your own energy-providing exercise program. Because not all causes of fatigue are purely physical, Chapter 9 will address emotional, psychological, and spiritual habits and activities that can steal your energy, and it will present techniques for changing these beliefs and behaviors. Chapter 10 will help you plan for the future.

At the end of most chapters (5, 6, 7, 8, and 9), the chapter's activities are broken down into three steps you can take to eventually reach the chapter's goal. You can take all of these steps at once, or you can take them one at a time over a period of many days, weeks, or months. That is up to you. But let me remind you that you may not achieve any results unless you go on to step two or even step three. Each chapter is followed by a "high-energy tip," suggesting some relatively simple information or activity that will help you along the way to high energy.

I have two friends who are marathoners. One is a world-class runner who finishes marathons in under two and one half hours. My other friend is somewhat short of world class. It takes him more like six hours to run the same 26.2-mile course. The world-class runner has told me he cannot understand how our slower friend can run for six hours. Our slow friend has said that if he had to run the marathon in under three hours, he simply would not be able to do it. Some people take slower, smaller steps, while others take faster, larger strides. Yet both eventually cover the same distance and get to the same finish line.

Though you may take longer to do it, and though you may stop to catch your breath and look back at how far you've come, I encourage you to cover the entire distance in my program. The idea is not to force yourself to do things you don't want to do, but to learn what works for you and what doesn't, to learn what gives you energy and what takes it away. Each step will give you a new tool with which to achieve that goal and attain high energy, optimal health, and peace.

HIGH-ENERGY TIP

A QUICK PICK-ME-UP WITHOUT CAFFEINE

Many of my patients who complain of fatigue drink coffee, tea, colas, and chocolate to boost their energy. Many of them claim they have read somewhere that the caffeine in these drinks really does boost alertness, stamina, and performance capacity.

Unfortunately, they are only half right. Caffeine does stimulate the body to produce more energy and higher performance in physical and mental tasks. But the effects are only for short-term use. Chronic use of caffeine depletes energy and brings about a drop in performance. In Chapter 4, I'll explain exactly why this happens. But for now, I want to make it clear early in the book that caffeine cannot be a regular "pick-me-up."

In tests with athletes, caffeine was found to boost performance and stamina. In a study performed by David Costill of Ball State University, 60 cyclists were each given 250 mg. of caffeine one hour before a long ride. Then, during the first 90 minutes of their ride, they were given 250 mg. of caffeine every 15 minutes. As a result, the cyclists' oxygen consumption rose 7.3 percent, but their perception of exertion remained the same. In other words, they got more work done but didn't feel as though they had exerted themselves any more than normally.

Basically, this study demonstrates that when a person who *doesn't normally use caffeine* does so prior to and during an athletic event, the caffeine helps the person do more work and perform better, and boosts stamina and endurance. However, *chronic* use of caffeine affects performance adversely. Several studies have demonstrated that chronic use of caffeine actually decreases performance, stamina, and endurance. Furthermore, many people are sensitive to caffeine and other elements in coffee, tea, cola, and chocolate. They may experience a phenomenal boost the first time they use caffeine, but then sink deeper into fatigue even more quickly as they continue to use it. Like all drugs, caffeine does not let the user off the hook easily. After stopping chronic use you may experience withdrawal symptoms— fatigue, irritability, lack of concentration, nervousness— that occur more and more frequently and more and more intensely.[1]

The belief that caffeine can create energy on a day-to-day basis is a myth. The practices in this book will help you achieve true high energy that will boost your performance, stamina, and endurance to levels you could never reach with caffeine.

Instead of caffeine drinks try these simple, natural "stimulants":

1. Take a shower. Alternate the water temperature between hot (not scalding) and cool two or three times. Finish with cool. This will stimulate you naturally.

2. Exercise. You don't have to run five miles. Just get up from your desk and stretch. Then try some brief calisthenics. Shake your arms. Wave them. Jump up and

down in place for 20 seconds or so. Take a short, brisk walk—but don't park yourself by the snack machine.

3. Relax. Many times we feel fatigued because our own nervous energy bottles up inside us and, unable to find release fast enough, tires us out. You may experience this if you have a big job in front of you, even if you are excited about the task. It sometimes helps to relax before you continue working. Just lie down or sit in a comfortable position, do some stretching exercises, take 10 deep, slow breaths—and feel your energy return.

4. The Vitamin C Drink described in Chapter 5 is an excellent, effective replacement for caffeine. It's easy to make and replaces a poor-quality beverage with a high-quality one. It increases energy and helps to promote bowel movements, two specific reasons why people use coffee.

5. Listen to music. Some music relaxes you. Some energizes you. Some makes you want to get up and dance. All three of these kinds of music can be useful at the right times. If you need to concentrate on a task, music that makes you want to dance may not be appropriate and may actually serve to drain your energy. However, music that relaxes you and allows you to let your natural inner energy come forth may be just what you need. What kind of music to use in a given situation is up to you. It depends on your tastes and your personal response to different kinds of music. Just keep in mind that music can accompany you on the path to high energy.

2

What's Your EQ?

"The real voyage of discovery
Consists not in seeking new landscapes
But in having new eyes."

MARCEL PROUST

If you came into my office and told me you were tired, I would want to find out why. I need to know what, exactly, is causing your fatigue before I can tell you what to do about it. I need to know your EQ, or Energy Quotient, which tells us how far you have to go before you have high energy.

So I would ask you a series of simple questions. From your answers I would be able to tell you what is making you tired and what you must do to get your energy back.

With the Energy Quotient Self-Test, you can do the same thing. The test reproduces the same process I go through to diagnose energy problems for my patients. You can write in the book, use a piece of scrap paper, or photocopy the test and write on that.

The test is divided into six parts. Your answers to each section will give you some clues as to what specific areas of your lifestyle and diet are causing your fatigue. (I don't want to bias your answers, so I am purposely not labeling the parts of the test with the names of the problems they diagnose.) For the first five parts, just answer the questions by writing down a number:

If your answer is NEVER, put down 0.

If your answer is SELDOM, put down 3.

If your answer is FREQUENTLY, put down 6.

If your answer is ALMOST ALWAYS, put down 9.

At the end of each section, add up your score and then go on to the next section.

SECTION 1

NEVER=0 SELDOM=3 FREQUENTLY=6 ALMOST ALWAYS=9

1. Are you consistently tired?   ———

2. Has a doctor ever told you that you are anemic? [Yes = 6; No = 0]   ———

3. Has a doctor ever told you that you have a low thyroid condition? [Yes = 6; No = 0] _____

4. Do you get dizzy if you stand or sit up quickly? _____

5. Do you feel short of breath? _____

6. Do you have a lot of trouble controlling your weight even when not overeating? _____

7. Do you feel cold when others around you don't? _____

Please total your score for Section 1:

SECTION 2

NEVER = 0   SELDOM = 3   FREQUENTLY = 6   ALMOST ALWAYS = 9

1. Do you wake up tired? _____

2. Do you feel weak or tired if you skip meals or don't eat for a long time? _____

3. Do you get headaches, become irritable or depressed if you skip meals or don't eat for a long time? _____

4. Does eating relieve your fatigue or other symptoms? _____

5. Do you get especially fatigued in the late morning or late afternoon? _____

Please total your score for Section 2:

SECTION 3

NEVER = 0   SELDOM = 3   FREQUENTLY = 6   ALMOST ALWAYS = 9

1. Do you feel tired or exhausted within two hours of eating a normal-sized meal? _____

2. Do you feel depressed, confused, or get a headache within two hours of eating a normal-sized meal? _____

3. Do you feel better if you don't eat for long periods of time? _____

4. Do you have a lot of indigestion, gas, or bloating in the intestines? _____

5. Do you have a persistent sinus drainage (runny nose) or postnasal drip? _____

6. Does exercise improve your feelings of exhaustion or fatigue? _____

Please total your score for Section 3: _____

SECTION 4

NEVER = 0   SELDOM = 3   FREQUENTLY = 6   ALMOST ALWAYS = 9

1. Do you work in an office at a desk-type job? _____

2. Do you get too tired to exercise? _____

3. Do you watch television two or more hours a day? _____

4. Is it a problem to work exercise into your schedule? _____

5. Do you need nine or more hours of sleep a night? _____

Please total your score for Section 4: _____

SECTION 5:

NEVER = 0   SELDOM = 3   FREQUENTLY = 6   ALMOST ALWAYS = 9

1. Do you worry? _____

2. Are you having trouble with a personal or professional relationship in your life? _____

3. Are you in financial difficulty?   ____

4. Is it hard for you to relax?   ____

5. Does life seem tedious and troublesome to deal with?   ____

   Please total your score for Section 5:

SECTION 6:

In this section, I'm going to ask you to do something a little different. I want you to look at the following series of graphs. These are graphs of daily energy ups and downs. Examine them carefully and select the graph that most resembles your own daily energy pattern.

Don't worry about choosing the "right" or "wrong" graph. There's no scoring for this part of the test. However, in the next step of the program, after we've scored your Energy Quotient Self-Test, we will look at these graphs again to get an idea what your selection means.

Graph 1

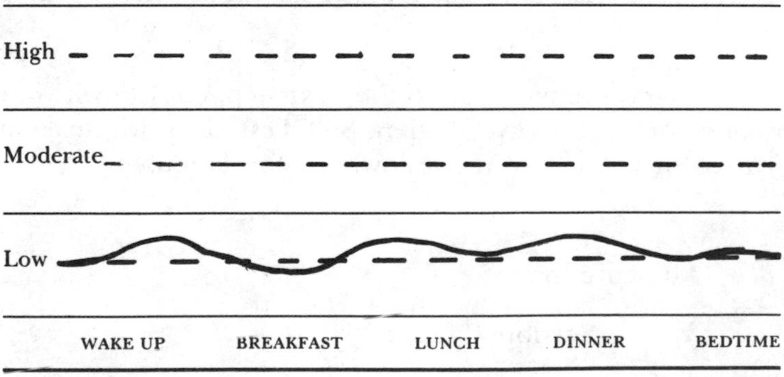

High – – — – — — – — – – — – — — – —

Moderate_ _ _ — — – — — – — – — – — – — —

Low —

WAKE UP   BREAKFAST   LUNCH   DINNER   BEDTIME

Graph 2

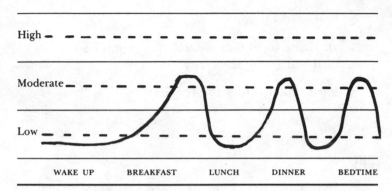

High

Moderate

Low

WAKE UP   BREAKFAST   LUNCH   DINNER   BEDTIME

Graph 3

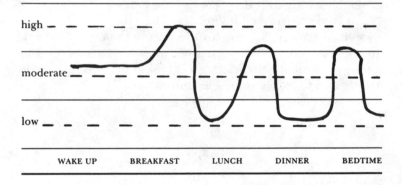

high

moderate

low

WAKE UP   BREAKFAST   LUNCH   DINNER   BEDTIME

Before we move on to the next step, copy down your scores for the Energy Quotient Self-Test. That will make it easier for you to use the following information.

My score for:

Section 1 _____

Section 2 _____

Section 3 _____

Section 4 _____

Section 5 _____

TOTAL SCORE _____

In the next section you'll find out what's causing your fatigue and what you can do to get high energy and keep it. As I explain the different sections of the test, bear in mind that if your score on a particular section of the test is 24 or higher, that section is significant for you. It will point you to a step in my program to which you should pay particular attention.

If you score 30 or higher in any section, that section is *very* significant. It is essential that you pay very close attention to that part of the program—otherwise you may not regain full energy and keep it.

Does this mean you should pay attention only to those steps in which you scored 24 or higher? Absolutely not! I can't emphasize strongly enough how important it is for you to read the entire book and have some knowledge of all the factors involved in energy. They really do interrelate. One or two may be causing most of your problems, but the others also play minor roles. You need to know what all the factors are so you don't miss any of them.

SECTION 1: BE ALERT FOR ANEMIA OR LOW THYROID

If you scored 24 or higher in Section 1, your fatigue is caused either by anemia or a low thyroid condition. I put this section first because you cannot treat these problems yourself; you must seek a physician's care. These problems are among the few causes of fatigue that conventional doctors are able to diagnose and treat successfully.

Question 1, Are you consistently tired? refers to what

is perhaps the most common symptom of anemia and low thyroid. Other forms of fatigue will usually—but not always —result in fatigue that comes and goes. If you have anemia, chances are good that you almost never feel very energetic. If you look ahead to the energy pattern graphs, you'll see that whereas the other problems produce energy patterns more like roller coasters, anemia and low thyroid produce a pattern (Graph 1) that looks more like a crawl across a dry desert lakebed.

Question 2, Has a doctor ever told you that you are anemic? seems rather obvious. But you would be surprised at the number of people who have been told by a physician that they are anemic, bought a bottle of iron supplements, and then proceeded to forget about the problem once the bottle was used up. Anemia can be caused by a temporary condition, but it can also be a chronic tendency that must be consistently treated.

Question 3, Has a doctor ever told you that you have a low thyroid condition? is included for the same reason. Low thyroid is usually a lifelong concern. If you've had it before, you probably still have it.

Question 4, Do you get dizzy if you stand or sit up quickly? refers to something doctors call "postural hypotension," which simply means that your blood pressure is low and that you get dizzy when you stand up because the blood supply to your brain is momentarily reduced. Low thyroid can cause this condition by keeping your blood pressure low. Anemia can bring it about by thinning your blood to the point where your brain's supply of oxygen is marginal, so that even the slight momentary drop in blood pressure caused by standing up will bring about dizziness.

Question 5, Do you feel short of breath? refers to a symptom that's common in anemia, although it can also be caused by asthma, allergies, and air pollution.

Question 6, Do you have a lot of trouble controlling your weight even when not overeating? refers to one of the classic symptoms of low thyroid. Most overweight people will admit, in moments of honesty, that they overeat. But some overweight people are at a genuine loss to explain why they can't lose weight. They claim that they gain weight even when they watch someone else eat! They lose weight only with great difficulty and gain it quite easily. Why? Because their low thyroid condition slows metabolism to the point where they do not require many calories to keep them going. Remember, they will also have a reduced level of activity because their energy fires are dimmed. With a slow metabolism using less fuel, there are more calories available for storage as fat.

Question 7, Do you feel cold when others around you don't? refers to another classic symptom of low thyroid. Our energy fires also help keep us warm. When the thyroid is low, the body's thermostat is also turned down a few notches, so the body's "furnace" burns at a slower, cooler rate. When your energy flame is low, your body produces less heat. You will tend to feel cold even when others do not. When everyone else is comfortable in a short-sleeved T-shirt, the low thyroid person will often need a long-sleeved sweater.

In Chapter 3, Do You Need to Go to the Doctor? I'll give you information about these problems to help you understand them, and explain how a doctor diagnoses and treats anemia and low thyroid. I'll tell you how to report your symptoms and ask the right questions in order to obtain the correct diagnosis and the best treatment. There are a few other medical problems that could be causing your fatigue, such as mononucleosis, an overwhelming *Candida albicans* yeast infection, or heavy-metal poisoning. Though these are relatively rare, if you suspect you are subject to any of these conditions, you should explore the possibilities further with your physician.

If your fatigue does require a physician's care, the usefulness of this book and program doesn't end here. Read the rest of the book to gain a better understanding of energy problems that you may still have once your anemia and low thyroid are corrected. As a matter of fact, it's quite common for people with anemia or low thyroid to suffer from one or

more of the other fatigue-causing problems as well—which
I'm sure the Self-Test has already told you.

SECTION 2: THE HYPOGLYCEMIA HYPOTHESIS

If you scored 24 or higher on Section 2 of the Self-Test,
your fatigue is caused by hypoglycemia, or low blood sugar.
Chances are you're using sweets, coffee, and cola to whip
energy out of your body. Your fatigue is caused by your
body's inability to balance blood sugar levels. Abnormally
rapid and severe blood sugar swings, stimulated by the foods
that satisfy your sweet tooth, are taking a heavy toll on your
adrenal glands, the glands most responsible for maintaining
your energy supply. Normally these glands, with the aid of
the liver and pancreas, would be able to keep your blood
sugar relatively level during the day. But the junk food you've
been eating has weakened your adrenal glands to the point
where this function is severely compromised. Weak adrenal
glands cannot secrete the hormones necessary to keep blood
sugar at levels sufficient for high energy.

Question 1, Do you wake up tired? is commonly an-
swered in the affirmative by people with low blood sugar.
Normally your blood sugar will be balanced between meals
throughout the day or night. In a low blood sugar condition,
however, that balance cannot be maintained. When you go
more than a few hours without eating, your blood sugar
drops. So while you're sleeping six, seven, eight, or more
hours, the drop in blood sugar can be quite steep. You wake
up tired.

Question 2, Do you feel weak or tired if you skip meals
or don't eat for a long time? refers to the same process. Your
body is unable to keep blood sugar on an even keel. Every
time you eat something, you get a temporary blood sugar
boost. But within a few hours, your blood sugar drops to the
fatigue level again.

Question 3, Do you get headaches, become irritable or depressed if you skip meals or don't eat for a long time? refers to different symptoms arising out of the same process described in the previous two questions. Of the body's organs, the brain has the most demanding continual requirement for blood sugar. When your blood sugar dips below a certain level, not only will you feel tired, but you may also experience mental symptoms. Headache, irritability, depression, and inability to concentrate are the most common.

Question 4, Does eating relieve your fatigue or symptoms? refers to the temporary boost in blood sugar you experience when you eat something.

Question 5, Do you get especially fatigued in the late morning or late afternoon? refers to the times of day when people with low blood sugar characteristically experience fatigue and other symptoms. These periods occur three or four hours after breakfast and lunch, just about the time when blood sugar drops.

Chapter 4 will explain how the adrenal glands work, how the wrong foods cause energy-draining blood sugar swings, and what you need to do to stop driving your adrenal glands into a state of almost constant fatigue. Chapter 4 will explain your problem, but you'll have to go beyond that and pay close attention to chapters 5, 6, and 7, too. Chapter 5 will help you break your addiction to energy-draining foods. Chapter 6 will tell you how to rebuild your diet from the ground up—and your energy supply at the same time. Chapter 7 will tell you what nutritional supplements and special healing foods you need to get more energy and keep it.

SECTION 3: SUSPECT A SENSITIVITY TO CERTAIN FOODS

If you scored 24 or higher on Section 3 of the Self-Test, your fatigue is probably caused by a food sensitivity. We frequently grow addicted to common foods and other substances because they give us a slight energy kick. Like the false stimulation sweets give the body, sensitivity-provoking foods only serve to weaken our adrenal glands. When the allergenic food enters the system, the adrenal glands are stimulated. Energy-producing and certain "well-being" hormones are secreted that effectively mask the actual effect the food is causing. Food sensitivities also directly affect the brain. They can make us depressed, confused, or mentally weary.

Question 1, Do you feel tired or exhausted within two hours of eating a normal-sized meal? and Question 2, Do you feel depressed, confused, or get a headache within two hours of eating a normal-sized meal? are meant to distinguish a food sensitivity reaction from low blood sugar. The fatigue and mental symptoms associated with low blood sugar usually occur more than two hours after eating. Of course, there is always a certain amount of overlapping. Many people actually have both conditions operating at the same time, to a degree, so we need additional information to narrow it down.

Question 3, Do you feel better if you don't eat for a long period of time? will supply that information. The person with low blood sugar rarely, if ever, feels better after skipping a meal. The person with a food sensitivity usually does feel better when a meal or two is skipped. Fasting allows the allergenic effects to be cleared out of the system.

Question 4, Do you have a lot of indigestion, gas, or bloating in the intestines? helps narrow the diagnosis further. People with food sensitivities usually have difficulty digesting their food completely. Partially digested food is passed through the bowel, causing upset, gas, bloating, and other symptoms.

Question 5, Do you have persistent sinus drainage (runny nose) or postnasal drip? refers to common allergic symptoms. Though food sensitivities do not arise out of quite the same biological mechanism as hay fever and other inhalant allergies, the symptoms are sometimes similar: the body may try to deal with the offending substance by producing mucus.

Another excellent clue for the diagnosis of food sensitivities is the "allergic shiner" look. Seen in both children and adults, a puffy, discolored area under both eyes, often accompanied by a slightly swollen upper nose, is a typical sign of allergic reactions.

Question 6, Does exercise improve your feelings of exhaustion or fatigue? refers to one method people may use, either consciously or unconsciously, to break through the fatigue caused by food sensitivities. A lot of the fatigue caused by food sensitivities may be only a *false perception* of fatigue caused by the sensitivity's direct effect on the brain. I often refer to this as "brain fog." The act of getting up and exercising, focusing the mind on action, often helps clear away that fog and allows available energy to be used. I must emphasize, however, that not all sensitivity-caused fatigue responds to this method. Food sensitivities can severely weaken the adrenal glands to the point where fatigue is physiologically real and the body simply cannot produce energy.

Chapters 4, 5, 6, and 7 are vital to regaining your energy. Chapter 4 will explain how food sensitivities attack the adrenal glands and weaken your ability to produce energy. Chapter 5 will tell you how to detoxify yourself, cleanse your body of fatigue-causing poisons, and avoid the foods that have been draining your energy. Chapter 6 will tell you what foods should go into your new high-energy diet. Chapter 7 will tell you what supplements to take for high energy.

SECTION 4: GET MOVING

If you scored 24 or higher on Section 4 of the Self-Test, your sedentary lifestyle is draining your energy. Without some form of regular exercise, all the systems of the body grow weak, including your energy-producing systems. Your cardiovascular system weakens and cannot deliver optimal levels of oxygen and fuel to the cells to burn for energy. Your digestive system becomes sluggish and the nutrients you need are not assimilated properly. Your muscles grow soft and flabby and simply can't do the work any more. Your endocrine system, which is responsible for controlling and stimulating the body through glandular secretions, including those of the adrenal glands, fails to stimulate the body into action.

Question 1, Do you work in an office at a desk-type job? refers to a common condition that prevents many people from getting enough exercise. People who work in physically demanding occupations, have a greater chance of getting enough exercise. But if you pilot a desk—or even a plane, for that matter—you are not doing what your body was made for: physical, aerobic work.

Question 2, Do you get too tired to exercise? refers to a common complaint of fatigued people. This is more than an excuse—it's usually true. The fatigued person who does not exercise *will* generally be too tired to do it. The point is that you must begin the cycle of building energy someplace. The energy you have lost by not using your body will not appear magically. You have to take the first step and "prime the pump."

Question 3, Do you watch television more than two hours a day? points out one of the most powerful energy drains in our society. Television acts like a soporific drug. It demands that the viewer sit or lie still and be passive. Such behavior is an energy drain pure and simple.

Question 4, Is it a problem to work exercise into your schedule? refers to a common excuse people give for not exercising. We always manage to make time for those things that are truly important to us. If having high energy is important enough to you, you will make the time to exercise. It won't mean that you'll get less done; you'll have so much more energy, you'll be able to get more work done in less time.

Question 5, Do you need nine or more hours of sleep a night? refers to a common symptom of nonexercising people. Physically fit people actually require less sleep than the nonfit. People who exercise tend to sleep better, too, less fitfully and more deeply and refreshingly.

If you want your energy back, you have to start a regular exercise program. You don't need to get into triathlon shape. You can pick a regular exercise or sport that you enjoy. Chapter 8 will tell you how to design an exercise program that will restore your energy while adding an enjoyable activity to your life.

SECTION 5: STEP AWAY FROM STRESS

If you scored 24 or higher on Section 5 of the Self-Test, then stress is taking a physical *and* metaphysical toll on your energy. Energy isn't just a matter of nutrients, muscles, glands, and biochemical balances. There are also mental, emotional, and even spiritual factors.

Question 1, Do you worry? is central to the diagnosis of energy-draining stress. Nothing drains more of your energy than worrying.

Question 2, Are you having trouble with a personal or professional relationship in your life? is straightforward. Such troubles use up a lot of your energy. The amount of energy drained is up to you.

Question 3, Are you in financial difficulty? is also straightforward. Few of us are independently wealthy, so our perception of our financial situation is important. If you believe that you are in financial difficulty, then that belief is draining your energy. You may believe that in order to have peace of mind, you need first to improve your financial situation. The truth is that you can have peace of mind right now—in fact, you may not be able to change your financial situation until you do so. Peace of mind is exactly that, a state of mind. It's not a state of the world or of the people around you, neither of which is under your control. It's a state of your own mind, a state you can control to a great extent.

Question 4, Is it hard for you to relax? refers to a situation that is common in a life filled with stress. When stress piles up on us, unless we try really hard to maintain our relaxation habits, those relaxing activities are often the first to be crowded out in favor of actions we believe will somehow get us out of the stress. The person in financial difficulty, for example, may feel that trading the morning exercise routine for more work time is a good bargain. It's not. The time spent in relaxation is more valuable as a reducer of stress than the extra work. A relaxed person is more confident and efficient and can do more work, and better work, than the anxious, stressed person.

Question 5, Does life seem tedious and troublesome to deal with? refers to your view of your life right now. If you view life as tedious and troublesome, chances are you are overstressed. One of the principles of energizing yourself from within is that belief creates experience. So if you believe that life is troublesome and tedious, then it probably is for you. You can learn the trick of exchanging that energy-draining belief for one that will allow you to see and experience life as an exciting, energizing adventure.

Energizing yourself from within is a matter of learning some basic techniques to help you live day to day without creating a lot of energy-draining stress and pain. Chapter 9 will teach you these techniques and get you started energizing yourself from within.

SECTION 6: YOUR PERSONAL ENERGY GRAPH

The last section in the Self-Test asked you to examine some graphs and pick the one that most closely resembles your personal energy patterns. Keep in mind that these graphs are idealizations. Your personal energy ups and downs may not be quite so clear. If yours resembles more than one graph, that's fine. Very few people suffer from only one energy problem. A mixture of daily energy patterns may result in a unique graph. Don't worry. At least now you have a picture of your personal energy pattern to compare with the normal, high-energy graph, which I will show you a little later.

First, take a look at these graphs below. They are the same as the graphs earlier in the chapter, except that here the energy problems that create such patterns are identified.

Graph 1
Anemia and/or Low Thyroid

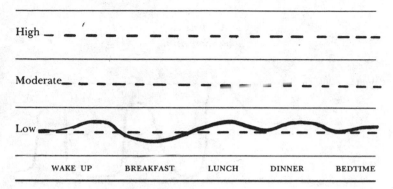

In this pattern, energy is always at low ebb. The body's energy fires seldom flare when you're suffering from low thyroid or anemia.

Graph 2
Hypoglycemia

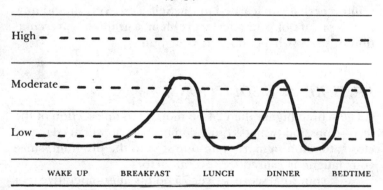

With hypoglycemia, energy tends to jump after meals and then fall about three hours later. You will rarely last more than four hours without experiencing fatigue and other symptoms.

Graph 3
Food Sensitivity

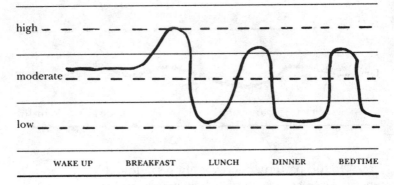

Food sensitivities tend to create an energy pattern that rises the further from a meal you get, only to drop sharply within two hours after eating.

Which graph looks most like yours? I asked you to compare your graph to these as a further diagnostic step to reinforce what you learned in the Self-Test. You should now have a picture of your energy problem, a graphic representation of what is happening in your body.

PICTURE THE SOLUTION

If you scored higher than 24 on more than one section of the test—a possibility that is quite common—you will need to pay close attention to more than one step in the program, since your fatigue is caused by several factors.

If your total score is over 75 points, it is imperative that you read the entire book and integrate all the therapeutic suggestions into your life to rebalance and heal yourself.

I cannot emphasize strongly enough that even if your entire problem appears solvable by following one step—say, for example, the exercise step—you will still pick up many important details by reading the entire book. As a physician I am interested in balancing all aspects of a person's life. We are not made up of distinct compartments. Our illnesses are usually not caused by single, isolated elements.

Just as you can look at a picture of your problem, you can also look at a picture of the solution to your problem. Here's a normal energy graph:

Graph 4
Normal

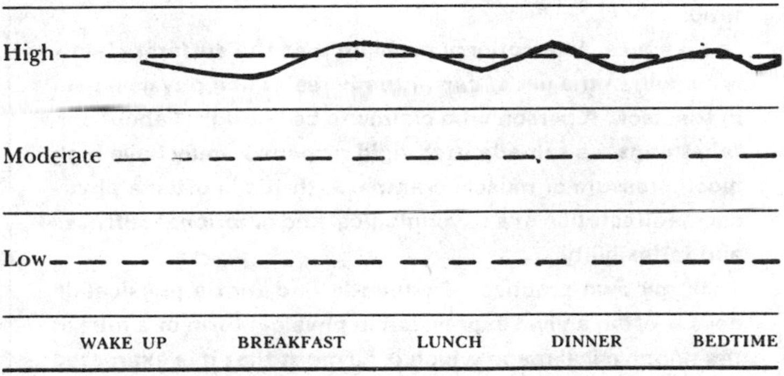

	WAKE UP	BREAKFAST	LUNCH	DINNER	BEDTIME
High —					
Moderate					
Low					

Your personal energy graph can look like this one. You now know what you have to do to get more energy and keep it. You're holding the map in your hands.

HIGH-ENERGY TIP

ARE YOU SICK AND TIRED BECAUSE YOU SAY YOU ARE?

All mental, emotional, and spiritual problems are eventually manifested or expressed physically. Our physical ills are sometimes an expression of some bottled-up emotion or strong feeling that we can't seem to let out in any other way. Sometimes we do express the feeling as a complaint without actually being aware that we're practicing a kind of self-hypnosis that is creating more distress while it draws from an already-existing problem. The commonly used phrase "sick and tired" comes to mind. People who habitually use this expression are truly sick and tired—the thing they claim to be sick and tired *of* is only secondary. Their reaction, however, is to mimic the emotional expression of disgust and frustration and be *physically* sick and tired.

Likewise, an emotional problem that the sufferer claims is "a pain in the neck" can actually result in a physical pain in the neck. A person who claims to be "uptight" about certain things— a self-admitted rigid response—may have high blood pressure or muscle cramps. Arthritis is often a physical manifestation of a psychological and emotional stiffness and inflexibility.

In my own practice I frequently find that a physical illness is often a vivid expression in physical form of a hurt in the nonphysical realm which is so great that it is expressed in the physical plane to call attention to itself and identify its need to be healed.

This is not an unreasonable explanation, since any kind of emotional upset—guilt, anger, sadness, worry, etc.—is not experienced only in the mind, emotions, or spirit. It will usually have a distinct physical component as well. When you are angry, the rage is not just a thought process. A large part of any upset is the particular physical manifestation that corresponds to that upset. The burst of energy, the pounding of your heart, the tension in your muscles, the physical feeling of being about to explode—these are what make anger what it is. You yell, scream, pound fists—or you don't, and instead store the energy inside you, where it eventually expresses itself later in some physical form in an attempt at release. Treating the physical ailment is the tip of the iceberg. It's the anger itself that needs to be released, and techniques are available to help accomplish this.

In the long run, a series of major and minor emotional upsets plays a distinct role in creating illness, reducing energy, and taking joy and happiness from your life. But you don't have to be a slave to these upsets. You can choose your response. In Chapter 9 I will describe in detail how to go about achieving freedom from energy-draining emotional attachments. For now, a tip to the wise is sufficient. Start taking inventory of your *statements* about the way you feel. Do you use phrases like "I'm sick and tired," or "I'm uptight," or "That's a pain in the neck"? Write down such expressions you use. Ask yourself if you are expressing these statements physically or emotionally as well as verbally. If you can recognize this activity, you will have taken a major step along the path to high energy.

Don't become emotionally dependent on or attached to your fatigue. Many people who have had a physical complaint for a while, be it fatigue or some degenerative illness, talk about it a lot, go to see a lot of doctors, and frequently talk about their problem to their friends and relatives. They seem to develop an emotional investment in their illness, as well as a literal investment of time, energy, and money. If you attach yourself to your fatigue, if it becomes part of your persona, then you're going to have a much more difficult time healing it. You must be willing to give it up. Look upon your fatigue as a temporary imbalance which you created and can resolve. This book will help you take responsibility and let go of fatigue.

3

Do You Need to Go to the Doctor?

"If you listen to the whispers
You'll never have to hear the screams."

LAZARIS, SPIRITUAL TEACHER

Are you consistently tired? Has a doctor ever told you that
you are anemic? Has a doctor ever told you that you have a
low thyroid condition? Do you get dizzy if you stand or sit up
quickly? Do you frequently feel short of breath? Do you have
a lot of trouble controlling your weight even when not overeating? Do you feel cold when others around you don't?

These, of course, are the questions I asked you on the
Energy Quotient Self-Test. You now know that if you scored
24 points or higher in Section 1 of that test, you probably
have anemia or a low-functioning thyroid gland. These problems should be diagnosed and treated by a physician. If your
thyroid is sluggish, it is absolutely necessary for you to have
it diagnosed and properly treated. Otherwise, none of your
efforts to regain energy will be ultimately successful. You
cannot ignore a low thyroid condition. Likewise, if you are
anemic, your body will always be at a disadvantage as you try
to regain high energy.

Anemia is a state in which your blood is deficient in its
ability to transport oxygen. Since almost all of the body's
metabolism depends on oxygen, any interference with the
body's ability to transport oxygen to the cells—and carbon
dioxide away from the cells—puts you in a very depressed
metabolic state. All of your organs—brain, muscles, heart,
etc.—will function at a lower level.

Because of the regular blood loss during their menstrual periods, women are more predisposed toward anemia
than are men. Unless there's an obvious disease, a serious
injury, or an illness in which blood is lost, few men are
anemic.

If menstrual blood loss is so heavy that the body has
difficulty making up for the loss during the four weeks between menstrual periods, anemia is the typical result. In
many young women, anemia is only borderline—enough to
drain their energy, but not enough to cause other symptoms.

If the blood is the transport system of the body, then
the thyroid is the gas pedal. It controls the basal metabolic
rate at which all systems in the body function. If the thyroid
is sluggish, everything in the body is slowed: energy, metabolism, digestion, and elimination. If thyroid function is normal, things run at the appropriate pace.

A low thyroid condition is typically not cured, though it can be controlled. A sluggish thyroid problem usually lasts throughout one's life. Taking good care of yourself, along the lines described in this book, will help you have more energy. And if your low thyroid condition is dealt with medically, you will be able to reap the full benefits of the high-energy program.

THE IRON-CLAD CASE FOR ANEMIA

If you suspect you are anemic, you should request a blood test from your doctor. To help the doctor understand what's going on, describe your symptoms, particularly those that have been revealed by the Energy Quotient Self-Test. This is a relatively simple diagnosis of a problem the doctor often sees.

To test for anemia, only one small tube of blood will be taken. From this, your doctor will test your red blood cell count, your hematocrit, and your hemoglobin. The red blood cell count is the actual number of red blood cells per cubic millimeter in your blood. The hematocrit measures what percentage of your blood cells are red, and the hemoglobin measures the amount of actual oxygen-carrying material in your blood. These are the three primary measurements of anemia.

If these tests do not reveal anemia, the doctor might then order an additional test, a blood iron level. Iron is required for the proper structure and functioning of the red blood cells. If your body does not have enough iron, it cannot make enough hemoglobin to carry oxygen. Iron-deficiency anemia is simple to correct by supplementing the diet with iron and vitamin C, which aids in iron assimilation.

There are other possible causes of anemia that the doctor can investigate. Anemia can stem from blood loss through injury, heavy menstrual periods, vomiting blood, ulcerative colitis, and inflammatory condition of the bowel. If you're suffering from one of these conditions, however, it will be obvious to you that you're losing blood. There are other forms of anemia, such as hemolytic anemia and pernicious anemia, in which the actual formation or maintenance of the red blood cells is faulty because of some metabolic defect. These anemias can also be successfully treated; however, further tests will be required to make the correct diagnosis. Keep in mind that these anemias are relatively rare. The chance that your fatigue is caused by some exotic anemia is extremely small.

A few words about heavy menstrual periods: Heavy menstrual periods frequently result from a woman's body being polluted with toxic elements—the result of years of poor care habits. The menstrual period then acts, in part, as a detoxifying process.

In my experience, once a woman performs a detoxification program and eliminates junk food and other toxic elements from her diet, the menstrual flow will be reduced in both amount and duration. Furthermore, cramps, uncomfortable feelings, emotional ups and downs, and many of the other so-called premenstrual symptoms will also improve. Finally, her anemia will be corrected. Chapter 5 on detoxification and desensitization will tell you how to rid your body of fatigue-causing poisons.

Another possible cause of anemia is sluggish or underfunctioning bone marrow. Red blood cells are primarily created in the marrow of the long bones. As long as the marrow is working properly, the number of red blood cells is usually adequate. But when the bone marrow is sluggish, red blood cells are not manufactured in sufficient quantity.

The thyroid often plays a role in sluggish bone marrow, since that gland controls the pace of all metabolism, including red blood cell formation. If thyroid function is low, the bone marrow is frequently sluggish, too. Starting out with a low thyroid condition, you can also end up with anemia.

One of the most frequent causes of sluggish bone marrow is the administration of heavy doses of antibiotics early in life. When there is a long history of antibiotics use—in someone who had a lot of ear infections or sore throats and tonsilitis as a child, for example, and therefore used antibiotics on a long-term basis—sluggish bone marrow frequently results in adulthood. Bone marrow output won't be terribly low—just low enough so that the person always runs a borderline anemia.

If you do have sluggish bone marrow, don't be concerned that you might need acute medical care. You may not even require medical treatment. You can often stimulate your bone marrow with moderate amounts of aerobic exercise. The body senses that it has to make more red blood cells in order to adapt and do the exercise.

Sometimes an "athletic" anemia occurs in a person who does a lot of long-distance aerobic training. Marathon runners, long-distance swimmers or cyclists can experience this form of low-grade anemia, which is actually an adaptation by the body: the fewer the red blood cells, the faster they pass through the capillaries. This is a very unique situation, occurring most commonly in people who are involved in long-distance training. If you have this form of anemia, you shouldn't be concerned, but you should have it checked out.

In most cases of anemia, the doctor prescribes an iron supplement. Some supplements may cause stomach upset and constipation. However, there are plenty of good-quality iron supplements (available from a nutritionally oriented physician or a health food store) that will not produce these symptoms. You can try a few different ones at the recommended doses until you find one that works well for you. You should remember to take your iron supplements with vitamin C, which aids in the assimilation of this mineral.

Anemia is not necessarily a lifelong problem, as low thyroid usually is. You may, however, need to take therapeutic doses of iron (more than 20 mg. per day) for an extended period of time until your anemia is corrected.

THE LOW-THYROID BLUES

In my experience, low thyroid is poorly diagnosed through blood tests; it is better diagnosed by the patient's description of the condition. Low thyroid has several classic symptoms, some major and some minor. The major symptom is a chronic low level of energy, producing a flat, low-level energy graph. Other major symptoms, as I have described, include putting on weight easily and losing weight only with difficulty, and feeling cold and needing to overdress when others do not.

The minor symptoms include moodiness, unwarranted feelings of depression or anxiety, hair loss (particularly in women), swollen ankles, and water retention.

Most of the causes of low thyroid are not very obvious. One cause is a deficiency of iodine in the diet. In many areas of the world the soil is low in iodine, and goiters and low thyroid function are common among the residents. The "Goiter Belt" in the Midwestern United States is an example. One of the reasons salt is iodized is to protect against this happening to large segments of the population.

We're not sure exactly how low thyroid conditions come about. Perhaps there is a genetic predisposition that renders the thyroid a weaker gland. We do know that low thyroid is much more frequently seen in women than in men, and that the condition frequently runs in families. If a woman's mother, grandmother, and sisters have low thyroid function, then she is likely to be subject to it, too. A history of frequent miscarriages is another sign of a low thyroid condition.

It's also possible that low thyroid conditions begin with a slight iodine deficiency or an inherited weakness and then, after poor self-care, the thyroid becomes sluggish and does not make the amount of hormone that it should.

If you have reason to believe your thyroid function is low, tell your doctor. The doctor may order a blood test called a thyroid panel. If the test reveals low thyroid function, the doctor will prescribe thyroid hormone.

It's possible that your blood tests will come out "normal," and your doctor will say something like this: "I sure thought you had low thyroid, but that's not it. Your tests are normal, your thyroid's OK." As far as the doctor is concerned, the issue of thyroid is forgotten.

It has been my experience, however, that if the doctor listens to the patient long and carefully enough, the diagnosis may become clear despite the test results. I don't understand why we often see normal thyroid levels in people who tell a definitive story of low thyroid function. But I am more inclined to trust my experience and the patient's story than I am to trust a simple blood test. When I asked a respected, older physician about this once, he said, "Are you treating the test or are you treating the person?"

Why are the blood levels normal even though the person experiences problems? It is possible that some particular toxin or autoimmune reaction in the body is blocking the hormone from having the effect it should be having on the body. Recent research has revealed the presence of particular antibodies that block thyroid hormone from achieving its effects. This is one good possible explanation for the normal test–low thyroid story phenomenon—and you might bring this possibility to your doctor's attention. Therapies are available to correct this condition.

There are two other methods of testing thyroid function. The first is the photomotogram, an old-fashioned machine used to measure the Achilles tendon reflex. Low thyroid is indicated by a slower-than-normal reflex.

The other method, which has become quite popular, is called axillary underarm basal temperatures. This method was developed by Dr. Broda Barnes, who spent many years studying the thyroid gland. Dr. Barnes felt that this temperature measurement was a very accurate way to measure the effect of the thyroid hormone in the body, that even if the blood level was normal, this method would show what kind of effect the hormone was having in the body.

The test is done at home, using a basal thermometer, which is calibrated more precisely than a regular thermometer. It's available at any pharmacy. Before you go to sleep each night, shake down the thermometer and leave it on the nightstand. When you wake up in the morning, roll over and tuck the thermometer in your armpit for ten minutes. Read your temperature and write it down. Do this seven days in a row.

It's best for women to start this on the second or third day of their menstrual period. Men, and women past menopause, can start any time.

The normal temperature range is between 97.8 and 98.2 degrees F. If you're in that range, it indicates your thyroid is working fine. If you are consistently below 97, if your temperatures are all in the 96s or lower, then there's a good chance your thyroid is low. You are a candidate for further testing or for replacement thyroid hormone. You may need to find a doctor who will work with you on the basal axillary temperature and low-dose trials with thyroid hormone.

When I prescribe thyroid hormone, I always warn that there are two common side effects. One is heart palpitations, a definitive sense of the heart beating in the chest. Another is a jittery, wired, nervous, edgy kind of energy. You will definitely feel as though you have more energy—but it doesn't feel good. If either of these symptoms occurs, you should stop taking thyroid, reduce your dosage, or try another type of thyroid hormone. I might add that natural thyroid preparations are much less likely than synthetics to cause side effects. This is one of the advantages of the natural hormone. One of the advantages of the synthetic, on the other hand, is that the doses are much more uniform. I prefer to use naturals, although synthetics work just fine as long as the doctor is watching the patient for side effects.

Once you begin taking thyroid hormone, it pretty much becomes a lifelong obligation. Nevertheless, if you do need to take thyroid hormone, don't feel that it's the same as taking a drug. Thyroid hormone is an orthomolecular substance—it belongs in the body. It has a biological basis for being there, and the body has natural pathways for dealing with it, since the body creates thyroid hormone on its own.

If you do need thyroid hormone, once you take it you'll feel like your life just got turned on. You'll notice that much of a difference in your energy supply—and in all the other symptoms I've talked about.

TRICIA'S TALE

I want to close with an anecdote that demonstrates the importance of incorporating all of the energy-building techniques in this book—even if you feel that correcting your anemia or low thyroid condition is all that's necessary. Tricia is a 37-year-old woman who runs her own business. She came to me complaining of memory loss, inability to think clearly, depression, severe asthma for which she was taking two prescription drugs, constant fatigue, lack of concentration, heart palpitations, stiff joints, an inescapable sense of weakness, bloating, and gas. Tricia had been diagnosed as having low thyroid and had been taking replacement hormone.

After talking to Tricia and finding out more about her life, it became clear that her adrenals were weak from all the stress she was experiencing. In addition to running her business, she had to take care of her husband and children. She was also eating a lot of junk food and foods to which she was sensitive.

Tricia needed just about every part of my program. She performed the detoxification and desensitization, and, as is usually the case, her energy level took a giant step up immediately. From then on, as she followed subsequent steps, including avoidance of allergenic foods and junk foods, supplementation of her diet, exercise, and stress reduction, her energy continued to build slowly and steadily. Her symptoms have cleared up. Tricia not only has plenty of energy, but she can concentrate, and digest her food. Her stiff joints, heart palpitations, and asthma are also gone. Tricia has high energy.

HIGH-ENERGY TIP

DON'T WATCH TOO MUCH TV

Just as the wrong foods can steal your energy, so can the wrong activities. What you do with your time, the people and things you surround yourself with, are all forms of nourishment. TV is often bad nourishment—it drains energy. Watching the screen removes you one step from actually living your own life. Not the least of the energy drain comes from the commercials, which disrupt your train of thought and contain all sorts of subliminal manipulations. Many of them urge you to eat the wrong foods and encourage you to adopt energy-draining habits.

Many people also have the "popcorn-at-the-movies" mentality: whenever they're watching anything on a screen, they have to be eating something. They generally exhibit poor nutritional choices in these situations and frequently choose allergenic or junk foods.

I'm not suggesting that you should throw away your TV set. There are often very fine educational and entertaining programs on television. But what too often happens is that people make a habit of turning the TV on and watching the best program they can find. If you plan your TV viewing the same way you plan your other activities, by deciding beforehand what you are going to do based on how it adds to your life, then you will waste less time and energy watching TV.

It's possible that you're spending too much time being a passive spectator, whether it's TV, movies, theater or sports that you're watching. Be a participator instead. Use that same amount of time to play a game or sport or learn a new craft, hobby, or musical instrument. Read a book and stimulate your own imagination, instead of just observing someone else's—who doesn't have your best interests at heart anyway. You'll find that you feel much more alive, have more self-respect, and feel good about what you've accomplished. You'll also build more energy instead of having it drained away.

4

Are the Wrong Foods Draining Your Vitality?

"Nature understands no jesting. She is always true, always serious, always severe. She is always right, and the errors and faults are always those of man. The man incapable of appreciating her she despises, and only to the apt, the pure and the true does she resign herself and reveal her secrets."

GOETHE

Energy is so vital to us that we often equate it with life itself. If we have more energy, we feel more alive. So we're willing to pay dearly for it—sometimes too dearly, when we buy more energy in ways that are ultimately harmful. Our bodies crave vitality. Unfortunately, not knowing how to give our body what it needs, we often look for the easy way out, the easy energy "fix." Coffee, cigarettes, sugar, amphetamines, cocaine, and other drugs, including alcohol (which is technically a depressant, but which often acts initially as a short-term stimulant) become the driving force that keeps our body machinery going. For a price, of course.

Many people think they have energy, but they're creating it in all the wrong ways. They use these artificially stimulating substances to whip their bodies into high gear. They're burning their energy candle at both ends.

Are you one of them? If you scored 24 points or higher on Section 2 of the Energy Quotient Self-Test, you very well may be.

THE ADRENAL GLAND–ENERGY CONNECTION

Most practitioners treat fatigue as if it were a digestive malfunction that prevents energy from being released from food. This approach is usually limited to prescribing a diet high in complex carbohydrates and vitamin supplements. There is nothing wrong with supplements or complex carbohydrates —I recommend them myself in subsequent chapters. However, there is more to energy than releasing energy from food. Treating fatigue as a nutritional deficiency ignores the heart of our energy supply mechanism—the adrenal glands —and so ignores the heart of the problem of fatigue.

Our energy fire is stoked by our adrenals, two thumb-sized glands nestled one on top of each kidney. The adrenal glands are part of the endocrine system. They produce and secrete hormones directly into the blood, hormones that help regulate a wide range of our body's normal functions, including energy metabolism and sexual function. Because each adrenal gland has an inner portion that secretes its own hormones more or less independently of the outer portion, the gland is sometimes called the "gland within a gland."

The outer layer, or cortex, synthesizes and secretes three classes of hormones. The first class is glucocorticoids, which play a role in the metabolism of protein, fat, nucleic acids, and especially carbohydrates. Cortisol, the best-known glucocorticoid, is required to maintain the liver's role in regulating blood sugar levels. Glucocorticoids also influence the immune and inflammatory response, wound healing, and the integrity of the muscles.

The second class of hormones secreted by the cortex is called mineralocorticoids, which influence sodium and potassium metabolism. The third class of cortical hormones is sex hormones, which are secreted by the cortex in addition to the gonads.

The interior portion of the adrenal, the medulla, secretes catecholamines, epinephrine and norepinephrine, which are central to the "fight or flight" response.

Actually, the best-known role of the adrenal glands is to set up the body for "fight or flight" during periods of stress. This is central to our understanding of how all kinds of stress affect our energy supply, since the body's basic fuel, blood sugar, is regulated largely by the adrenal glands. When the body is challenged by a stress, hormones pumped out by the adrenals put the body's energy system on alert. That "alert" includes raising the blood sugar to supply energy for fight or flight. As long as the challenge is perceived as a stress or a potential danger, there will be plenty of energy for quick action. As far as the adrenals are concerned, it doesn't matter if the challenge is a ferocious wild beast, a cantankerous boss, a traffic jam, a job that *must* get done—or a foreign substance invading the body to "stimulate" it.

All the stresses in life—physical, chemical, emotional, mental—put the adrenals to work. When the stresses are excessive and of long duration, the adrenal glands are slowly but surely taxed. If your nutrition is not adequate and your other lifestyle habits do not support the health of your endocrine system, the adrenals become semiexhausted and don't secrete your personal, normally adequate levels of hormones, especially adrenaline and the corticosteroids—unless, of course, they're forced into it by either a stimulant or an exaggerated stress. Furthermore, your adrenal fatigue will most likely not show up on conventional medical tests. Laboratory tests of blood and urine for adrenal hormone levels will usually not turn out abnormal except in cases of severe adrenal exhaustion or total failure.

If you whip tired horses—or tired adrenal glands—they'll run faster for a stretch. But in the long run, they just keep getting more and more tired. Eventually, the glands are able to stimulate very little natural energy in the body on their own.

That is why if you're a regular user of stimulants, you will most likely wake up tired, even after going to bed exhausted and sleeping 9 or 10 hours. Your adrenal glands cannot recover overnight from days, weeks, months, and years of artificial stimulation.

Caffeine, nicotine, and other stimulant drugs all have their own characteristic stimulating reactions in the body. But they also exert a "pick-me-up" effect directly on the adrenals in addition to whatever effect they have on the brain and central nervous system. When the craving for energy comes and one or more of these drugs is consumed, the adrenals are also driven to stimulate the body into a temporary energy state.

The length of the crave-consume-pickup-letdown cycle varies with the stimulant and the individual. Heavily dependent persons may require pickup doses every 10 minutes in the case of cigarettes, or every hour or so in the case of coffee.

Sugar, the "stimulant of choice" in our society, is not technically classified as such. But the rapidly available caloric energy provided by sugar does have a stimulating effect—for an hour or so. Then the high levels of sugar in the bloodstream cause the pancreas to secrete the hormone insulin, which enables two things to happen. First of all, blood sugar is taken up by the cells until they're practically saturated. There is an initial "rush" of fuel available for the cells to use. But then the second thing happens: since the cells have absorbed all the blood sugar they can possibly use at one time, the remainder goes into storage. The storage form of blood sugar is glycogen, which is deposited in the muscles and liver. The initial rapid pouring of sugar into the bloodstream calls forth such a profuse flow of insulin, that the blood sugar quickly disappears from the bloodstream into the muscles and liver.

The result is low blood sugar, along with fatigue, anxiety, depression, headache, and a host of other potential symptoms—and a craving for more quick energy. That quick energy can come from another dose of sugar—another soft drink or candy bar or even another full meal.

The craving is a natural protective mechanism for the brain, which depends on adequate blood sugar levels for proper function. Here, however, this basic regulatory balancing function of the body is abused into malfunctioning. The body is tricked into craving more and more sugar just to maintain normal levels of energy.

If you scored high on questions 2, 3, 4, and 5 of Section 2 of the Energy Quotient Self-Test, then you are experiencing this disruption of your body's energy metabolism. Your body's ability to balance blood sugar levels throughout the day is weakened to the point where it cannot be depended upon to keep blood sugar at normal levels. This is often referred to as hypoglycemia or chronic low blood sugar (although the term has been widely abused in recent years).

So you get tired. You may not even feel hungry at first. Your first symptom may be a temper tantrum, since your brain reacts to the low blood sugar first. If you don't eat soon, you'll start to feel weak. You may get a headache, feel lightheaded, or express a wide variety of mental and physical symptoms. The actual sensation of an empty stomach may be the last symptom—if you ever get that far without eating something.

Of course, the energy you need is already stored in your body in a readily available form. All the excess sugar and carbohydrates you've eaten have been stored as glycogen in your muscles and liver. But all that energy must be *mobilized* before it can be used. The glands that are mainly responsible for mobilizing energy stored in the muscles and liver are the adrenal glands—if only they can be persuaded to secrete the right hormones to liberate it. (The pancreas is also responsible. But your high-sugar diet has already put the pancreas in a state of near-exhaustion.)

The adrenals *can* be persuaded. One way is by exercising. Athletes and other people who are conditioned and fit have developed a good working relationship with their adrenal glands. So the glands cooperate when there is a need for energy.

Those of us who are not conditioned need some form of stress to persuade our adrenals to liberate some energy for us. So we fall back on a stimulant chemical: coffee, drugs, sugar, cigarettes, alcohol, or (as we'll learn more about in the next section of this chapter) some common substance to which we are simultaneously addicted and sensitive. All of these further tax the adrenals, making it even more difficult to call up our normal energy reserves.

A vicious cycle develops of fatigue and cravings for whatever will force the poor adrenal glands to give up some energy. And the ever-weakening adrenals will only exaggerate the low blood sugar energy dives and subsequent cravings. You will energize yourself, sure enough—but you will also fall into the trap of a compulsive appetite for the adrenaline rush of energy and for whatever stress will give you the hormone-energy "hit" you need. You may even create emotional stresses or upsetting experiences in your life to stimulate the flow of adrenaline.

You can conquer your cravings. You can rebuild your adrenal glands and create true high energy for yourself. What you must do, first, is to break free of all of the substances that keep hammering away at your adrenals. In the next chapter, you will learn how to detoxify your body and break clean of the substances that are stealing your energy.

First, however, we'll look at how some of your favorite, everyday foods can be draining your energy.

BEWARE YOUR FAVORITE FOODS

Even if you never touch a cigarette, coffee, or sugar, you may still be slowly overstimulating the life out of your adrenal glands. The offending substance may be a common food. In fact, it's quite likely one of your *favorite* foods.

If you scored 24 points or higher on Section 3 of the Energy Quotient Self-Test, your sensitivities to certain common foods may be giving you the same kind of energy jolt that a stimulant like sugar or coffee can—a jolt that picks you up temporarily before throwing you back down, a jolt that seriously disrupts your body's ability to feel energetic.

Just like drugs and other stimulants, food sensitivities have direct toxic effects on organs other than the adrenal glands. These direct effects are mediated through the immune system, which is involved in the intricate mechanism that actually produces the sensitivity reaction. One of the primary target organs for a food sensitivity is the brain. So even if your adrenals start out relatively healthy, a food sensitivity can, through its direct action on the brain cells, cause you to feel fatigued. Your brain is caught in the cross fire. It's directly affected by the food sensitivity, and also by the disturbance produced in the body's natural blood sugar balancing mechanism. But it also comes to crave the stimulation that occurs when the adrenal glands are jolted by the invader.

If you ever feel tired within a short time after eating, it's a sure giveaway of a food sensitivity. Eating a meal should not make you tired, depressed, confused, or give you a headache within two hours, unless there's something in that meal that directly affects your brain.

Are you crazy about dairy products, bread, chocolate, or corn chips? Is there any food you just can't seem to get enough of? If you know it's in the house, you keep getting pulled back into the kitchen for a little bit more—until you've polished it off. And even then you're left with a craving for it.

Most likely, you have an allergy—or, more precisely, a food sensitivity—to this substance. You are, in biochemical fact, addicted to the *effect* it has on your body. You also crave the offending substance because, as with any true addiction, having more of the substance relieves the symptoms of withdrawal—the main symptom being fatigue.

The addiction is biochemically mild, compared to the stranglehold exerted on the system by such drugs as alcohol, coffee, and nicotine. But the appetite, the compulsive pull toward the substance, whether it's chocolate chip cookies, corn chips, or milk, is at heart the same. Biochemically, only the strength of the compulsion and the degree of damage the substance inflicts on body and mind separate the "chocoholic," the "cornoholic," or the "milkoholic" from the alcoholic.

FOOD SENSITIVITY: THE MASKED ENERGY THIEF

Food sensitivity was first described several years ago by H.J. Rinkel, M.D., as a "masked food allergy." Rinkel called it "masked" because the victim doesn't suspect that he or she has this biochemical sensitivity. The person may actually seek out the offending food as often as necessary *to feel good.* In some cases the food causes headaches, excess mucus secretion, irritable bowel symptoms, decreased vitality, or even chest pain. Rinkel proved that a food sensitivity could be revealed by removing the food from a person's diet for four days and then reintroducing it; the food would then provoke an obvious reaction. He showed that continual consumption of the food would produce the masking effect and decrease the person's ability to notice the effects of the sensitivity. Because of the delay in onset of symptoms, the victim rarely makes the connection between the cause and the symptoms.

Why is there often a delay between eating the food and experiencing the negative effect? The differences between food sensitivity and the classical model of allergy can help explain this.

In the classic model of allergy, when the offending substance enters the body—whether through the digestive tract, lungs, or some other way—the body reacts, usually *immediately,* with an immunological response. The adrenal glands, ever on the alert against stresses of all kinds, also go into action.

We learned earlier in this chapter how the adrenal glands respond to a challenge or danger by pumping out hormones that provide energy to enable us either to meet the challenge head-on or to run away. But there's another part to the adrenal response. The adrenals also secrete cortical hormones that help to *control* many inflammatory reactions of the body to potentially harmful substances, including foods and chemicals to which we happen to be allergic or sensitive. These hormones serve to check the violence of the body's reaction to the invader. If this control mechanism breaks down, if the adrenals cannot secrete enough hormones, or if the assaulting substance is simply too violent an offender, the allergic reaction can be intense and life-threatening.

To neutralize such a reaction and save the victim's life, doctors inject strong, concentrated doses of the very same hormones the adrenal glands secrete to stem the violence of the response: corticosteroids and/or epinephrine.

During the allergic reaction, the offending substance is clearly perceived by the mind as well as by the body as an unwanted invader. In this type of allergic reaction, the body's response is immediate. The symptoms are predictable; they are usually mediated by histamine, and desensitizing injections are effective as a treatment. Most people subject to this type of allergic reactions are already aware of what particular foods or substances cause them. After scratching through a bout with the hives, having a runny nose all day, or having to be rushed to the doctor or even the hospital, the allergic person will not want to as much as *look* at the offending substance for a long, long time.

A different response occurs with food sensitivity. If you have an active sensitivity to wheat, for example, chances are good that if you don't get your wheat "fix" on a regular basis, you'll feel tired, dull, and lackadaisical from lack of the adrenal stimulation and the withdrawal effects. Depending on where you are on the cycle of withdrawal-craving, you will experience discomfort ranging from mild or barely perceptible all the way to violent withdrawal symptoms. The symptoms, usually delayed in onset, may include headaches, muscle spasms, depression, anxiety, mood swings, irritability, muscle cramps, joint stiffness and pain, itching, nausea, irritable bowel problems, and the most common symptom of this mechanism: fatigue. You may also experience some symptoms similar to classic allergy: runny nose, indigestion, bloating. In some cases, your symptoms may feel like a hangover. Many people stay at this level and "learn to live with" the lack of energy and uncomfortable symptoms and the addictions for most of their lives.

What is happening when you experience these symptoms is simply this: your adrenals stop secreting enough of the masking hormones and you then feel the direct toxic effect of the offending substance.

With food sensitivities, the body still treats the offending food or substance as a threat, so the response is still mediated through the adrenal glands, which prepare you for action by squeezing out hormones that give your energy and strength-producing systems a boost.

However, in addition, the cortical hormones that numb the effect of the body's inflammatory response to the invader are also secreted. So the single sensitivity-offending substance produces both energy and a subtle, hormonally narcotized feeling of well-being.

You get an adrenaline "high," and often, when your adrenals are not quite strong enough to suppress the direct sensitivity reaction, you will also feel some of the uncomfortable symptoms. You come to identify the high—but not necessarily the discomfort—with the food you've just eaten. You call it a chocolate high. A sugar high. A corn chip high.

Your appetite and craving for the food are biochemically rooted in the body and in the mind's craving for the adrenal kick that occurs when the offending substance enters the body—and in the body's desire to avoid the discomfort of the inevitable withdrawal symptoms.

Of course, the cycle rarely remains at the same level of intensity. A craving that began below the perceptual level will often rise to the all-too-obvious level. At some point you will reach your allergic threshold, where your immune system is overwhelmed by the continual insult of the offending allergens. Your adrenals are stressed and weakened to the point where their ability to suppress reactions grows less and less. You will then become more and more aware of both the initial reactions and the withdrawal symptoms—which may actually occur very close together if the adrenals are so weak that they cannot provide enough masking hormones for even a brief respite.

As your symptoms intensify, so will your cravings. Though you won't know why you can't seem to control the craving, you'll know precisely how awful you'll feel if you don't get what you want. It won't really be a matter of *want*, but *need*. You may be confused at this point. You may crave many substances, any or all of which may be contributing to your addiction. You may totally fail to recognize the pattern of addiction-withdrawal.

To use an extreme example, alcoholism can be explained as a food sensitivity in which the alcoholic craves more of the offending substance, which may be the ethanol (alcohol) or another component of his favorite drink. For a vodka drinker, it could be potatoes. For a whiskey drinker, it could be rye. For a beer drinker, hops, barley, or yeast. A wine drinker may really be craving the grapes or the yeast, not just the alcohol.

The alcoholic will maintain his craving for alcohol even if he stays "dry," by consuming nonalcoholic forms of the sensitivity-offending ingredients. A vodka drinker may go on potato chip binges, for example. The alcoholic will perceive this as a craving for alcohol and will not be able to understand why he still feels the craving even when he hasn't had any alcohol for a long time. By provoking a direct cellular sensitivity, these substances, even though they contain no alcohol, will keep the addiction-withdrawal mechanism intact and produce symptoms very similar to a "drunk" or hangover. This is actually the mechanism which explains the so-called "dry drunk" that many sober alcoholics complain of.

Frequently, alcoholics also satisfy—and thus maintain —the craving for alcohol by developing addictions to "acceptable" drugs like coffee, sugar, and cigarettes. These drugs not only tax the already tired adrenal glands and keep the addiction-withdrawal mechanism very much alive, they also keep the addict's body and life in a constant state of confusion.

The process is the same if you're addicted to chocolate, coffee, wheat, or corn. As the cycles of stimulation-withdrawal-craving intensify, as the compulsion begins to rule your life and drive you to the offending substance more and more, the adrenals will be taxed and weakened. The weaker they get, the less able they'll be to suppress the direct reactions and the withdrawal symptoms. Your compulsion will grow stronger because the withdrawal symptoms will come sooner and more intensely. Blood sugar is often low because the adrenals cannot stimulate the release of stored blood sugar from the liver and muscles. So you get tired. You crave whatever will whip your adrenals and give you some energy. And that may be either the sensitivity-offending substance, a stimulant, sugar, or a stressful emotional challenge.

There is really no limit to how far down the spiral of health and disease your weak adrenals can let you plummet. The weaker your adrenals get, the weaker your immune system gets. You will be left more vulnerable to all diseases, and the likelihood of developing additional food sensitivities and allergies will increase. You may even develop serious environmental allergies to common natural and artificial substances. Your life may be ruled not only by fatigue and by cravings, but by fear of the likelihood of running into any one of hundreds of common substances.

To free yourself from the compulsive appetite, the urge to consume the addicting substance, and reverse your progress on the health spiral, it's necessary to free the adrenals from the withdrawal-craving cycle and rebuild them.

The next chapter will show you how to begin to liberate, rebuild, and energize yourself from the inside out.

HIGH-ENERGY TIP

THE MYTH OF HYPERACTIVITY

There are two theories about hyperactive children. One, the conventional medical theory, states that some biological fault in the child's metabolism or nervous system results in the child's having *too much energy* and, therefore, being hyperactive. That is a straightforward theory that appears to make sense, until we take conventional medicine one step further and look at the treatment recommended for hyperactive children; Ritalin, a drug that is known as a *stimulant!*

How do conventional doctors explain why a stimulant is given to a child who supposedly has an excess of internal stimulation? They say that in children the drug has a paradoxical effect and acts as a sedative.

My next question is, at what point does a child reach the age at which the drug begins to act as a stimulant again? Where is the crossover point?

The conventional theory of hyperactive children doesn't stand the test of common sense.

The alternative theory of hyperactivity states that allergies to certain foods and artificial chemicals added to foods cause the sensitive child to be hyperactive. This is the Feingold Theory developed by Benjamin Feingold, M.D. This makes much more sense, and it seems to work in practice in many cases. My own theory follows along the same general line, with a few additions.

I believe that food sensitivities, allergies, and fatigue are at the heart of hyperactivity. Watch children's behavior shortly after they eat. If they become hyperactive or aggressive, this is a cardinal sign that food sensitivities are affecting their brains and directly producing a hyperactive state.

Schoolteachers frequently note that children's behavior is more hyperactive or their ability to pay attention is worse after a meal or snack. However, I do not believe in many cases that the internal biological effect of the sensitivities is to stimulate the child to hyperactivity. Rather, the child reacts just as an adult—with fatigue, brain fog, and lack of concentration. Quite often, then, hyperactivity is actually a low-energy state.

Anyone who's spent any time around children knows that when children get sleepy and groggy, they often react to the internal feeling with a kind of nervous explosion of energy; they get "cranky." They fidget, they fuss, they run around wildly—until they literally drop. If the child is suffering from the effects of a food sensitivity, or is for any reason in a naturally low energy state, he may fidget and fuss for hours. What is actually happening is that the child feels so tired, he or she creates a false nervous energy in order to feel any energy at all.

Adults often react the same way to food sensitivities and fatigue. They excite themselves into a state of nervous energy in a frantic attempt to push away the veil of brain fog and tiredness brought on by their food sensitivity reaction. Sometimes they use stimulants to do the job—in the same way that conventional medical doctors prescribe stimulants to hyperactive children. The stimulant does not have a reverse effect. It does just what it was made to do, stimulate. What really happens in hyperactive children is that the stimulant raises their base level of energy to the point where they can feel it, so they have no need to artificially create nervous energy and can then calm down. Since both childhood and adult hyperactivity are often just different expressions of a state of fatigue, these situations can be greatly improved or eliminated by following the program in this book.

5

Identify and Eliminate the Culprit Foods

"That which we are, we are,
And if we are ever to be
Any better, now is the time to begin."

ALFRED LORD TENNYSON

Sam was a 42-year-old man suffering from hay fever, frequent frontal sinus headaches, irritability, emotional ups and downs, manic-depressive behavior, dizziness, trouble focusing his eyes, excess sweating, spells of feeling faint, poor attention span, and tremendous feelings of fatigue. These symptoms were always worse after meals.

Sam knew he had food sensitivities to raw onions and bell peppers. I suspected he also had some sensitivities he didn't know about, so I instructed him in detox and desensitization. Four days into the program, Sam experienced a profound relief of symptoms. His sinusitis cleared up, the headaches went away, his emotional state balanced out, and his energy returned. Sam took a giant step forward during the detox and desensitization and then continued to make steady progress over the next few weeks. All of his symptoms faded away and his energy continued to increase. To this day he remains in very good health, with plenty of energy. For Sam, the detox and desensitization were the turning point in his life.

Karen was a 30-year-old woman whose major complaints were tiredness and inability to control her weight. She had a lot of responsibility and pressure at work, was always very tired after meals, and experienced frequent indigestion. Karen drank a cup of coffee a day and smoked a pack of cigarettes. She was too tired to exercise regularly.

During the detox-desensitization program, she reported renewed energy, which continued to build as long as she stayed away from junk food, stimulants, and her sensitive foods. Karen later agreed that it was a turning point in her life. Two months after she first came in, Karen reported that she had given up cigarettes, felt she had all the energy she wanted, and had become a regular exerciser. Moreover, she had lost 18 pounds. This was a particular triumph for her, since she had always feared that her weight would skyrocket if she stopped smoking.

This chapter can be a turning point in your life, too. In these next pages you are going to learn how to gain renewed vigor and confidence in your own energies. You're going to give yourself a chance to feel your energy begin to grow and

keep on growing . . . until you have plenty for all you want
to accomplish in life.

So say goodbye to the tired, irritable, self-limiting you
and prepare to meet a stimulating, eager, confident, optimis-
tic, fulfilled self—with energy to spare.

What kind of sacrifice must you make to bring about
this change? Surprisingly, just a little sincere effort and a
little patience. My detox-desensitization program is easy to
follow, virtually pain-free, and—with all the built-in rewards
it brings each step of the way—requires less willpower than
the average reducing diet. When you give your body what it
wants and needs to thrive, your body, mind, and spirit soar.
So enter into this program with a sense of adventure and
excitement, not sacrifice.

DETOXIFICATION—THE FIRST STEP

Your first crucial step is analagous to a spring house cleaning.
You'll be cleansing your body of years of accumulated toxins
from junk foods, chemicals, and drugs taken in both willfully
and involuntarily. You'll also be freeing yourself from your
food allergies and sensitivities at the same time. In addition,
it is not unlikely that you will find yourself undergoing an
emotional purge as well. The result of all this will be a more
physically and emotionally clear person, ready to go forward
effectively and able to read the subtle signs of progress more
astutely.

This is only the first step of several that you'll take as
you raise yourself higher and higher above the average degree of health, the normal ration of energy, to where you
belong, at the peak of optimal health.

For most of your life, chances are you've been eating
the wrong things, letting the wrong chemicals into your body
(drugs, cigarettes, alcohol), perhaps not breathing the cleanest air, not exercising enough, and enduring an unhealthy
load of stress. All of this abuse encourages unwanted chemicals and wastes—we'll call them toxins—to build up gradually in your body. These toxins are like grits of sand in a ball

bearing. They interfere with smooth operation of bodily functions and noticeably impair the body's ability to work effectively and efficiently. The detoxification is simply an attempt to clear a lot of this metabolic "garbage" out of the liver, bowel, and other tissues where it has accumulated.

The detoxification, or purification of body tissues from the effects of accumulated poisonous wastes, is achieved through modified fasting. Don't be scared off by that word. I don't believe that fasting should represent a hardship for you, either real or imagined, so I am recommending not a total fast, but a modified fast. You'll be allowed some caloric intake, in the form of diluted fruit and vegetable juices. And since I can't counsel each of you through your fast the way I do my patients, I have shortened the seven-day detox program I use in my office to just three days. Of course, many of you will feel so good on this modified fast that you'll want to carry it beyond the minimum three days.

The nice thing about the detox is that it accomplishes two objectives at the same time. It serves not only as a detoxification but also as a desensitization. I've talked about the concept of food sensitivities, where they come from, and how unhealthy, energy-stealing dependencies arise out of them. The modified fast will not only eliminate toxic body wastes but also desensitize you and break the stranglehold of food allergies and sensitivities.

It typically takes three to five days after the problem foods are withdrawn for the desensitization to take place. At the point when the body comes out from under the effects of these sensitizing foods, a remarkable thing happens for most people: they experience a tremendous surge of energy, a feeling of well-being, a renewed capacity for clearheadedness and creative thinking.

Many books on fasting have referred to this so-called "faster's high." I would like to offer an explanation for this feeling of increased energy, clarity, and creativity. I believe the high comes from the desensitization to food sensitivities that most people don't even know they have. When the fast goes on long enough (usually three days or more), this desensitization to offending dietary elements occurs.

As I've mentioned before, the brain is the primary target organ for sensitivities and is affected in a variety of subtle ways. So when you desensitize and remove the offending allergens, the brain will be the first organ to benefit. Not only will you become more clearheaded and creative, you'll also lose the deceptive perception of fatigue. Considering that you've been drastically restricting your intake of food energy during the modified fast, this perception of increased energy is all the more remarkable.

Not everyone will experience this faster's high to the same degree. Some people are healthy enough to shed the severe food sensitivities that had cast a veil over their brain, and then experience a remarkable turnaround of fatigue into energy. More common are the people whose adrenal glands have, over the years, taken such a beating that they usually don't experience a physiological high until their adrenals have been restored to health. Their highs come across as "mediums." They'll feel a degree better just from removing the insult of the offending foods, but they're not home free yet.

I'd like you to keep in mind a very important point: It took you many years to run down your adrenal glands, and you're not going to undo the damage overnight. So be patient with yourself. The detox-desensitization is meant as a preliminary maneuver to set the stage for rebuilding. The return of your full energy supply will usually not be accomplished in three days, five days, or a week. Any faster's high you experience is no doubt a blessing. Think of it as a sample of the wonderful feelings of power and vigor that will be coming to you more and more as you proceed in rebuilding your energy supply.

Granted, many of us at times feel uncomfortable with new experiences. We stifle ourselves with this or that excuse for not taking positive action: "It's too much trouble," or "I could never do that." Over 2,000 people have gone through this detox program under my personal guidance (most of them doing a longer program than the one described here) and have handled it without difficulty. You will, too. Once you're into it, the most powerful new feeling you'll have to get used to is the rebirth of your vitality.

Adopting a positive attitude is not only helpful in overcoming initial resistance to something new, it will actually shape your perception of the experience. If you're confident going into this detox that it will be a positive, healing experience, then it will be just that. If you go into it with fear and dread, dwelling on the negative and expecting the worst, then your mind will oblige you by exaggerating minor complaints and downplaying signs of recovery, including your renewed energy.

It's true that you may experience some ups and downs during the first three days. As you go into the detox, your body will seize on this break in its usual digestive and metabolic chores to start pulling toxic wastes out of the tissues and fat where they've been stored, in preparation for removing them. But until the liver, kidneys, lungs, skin, and other organ systems can work all the toxins out of the body, these substances will be floating around in your bloodstream. The toxins will have some effect, and you may feel a little discomfort.

It's not uncommon to go through some emotional vacillations, minor aches and pains, low-grade headaches, or light-headedness the first couple of days. On the other hand, many people breeze through the detox feeling wonderful the whole time. If you do have some uncomfortable symptoms, please hang in there! Now that you know exactly what is going on, there is no need to let fear make it worse than it really is. If you are hesitant to attempt this without direct supervision, or if you have any serious medical problems, then find a qualified, experienced physician to guide you through it.

In addition, women can increase the benefits of the detox if they time it to coincide with their menstrual periods. Menstruation adds one more channel for the elimination of toxins. Henry Bieler, M.D., a pioneer in the area of detoxification, believed that one of the reasons women live longer than men (on the average) is that menstruation gives them monthly opportunities to detoxify themselves for 30 to 40 years of their lives.

Of course, in some women, the toxic load can be so great that every month brings on an avalanche of distressing symptoms. Susan, age 33, was such a woman. Susan complained of fatigue and rather severe premenstrual syndrome (PMS). Within three to five days before her period, she would experience abdominal bloating, cramping, violent mood swings, and wide fluctuations in her weight. The detox-desensitization freed Susan from this monthly horror. She also gave up junk food and sugar, took the recommended supplements, and maintained a regular exercise program. Her PMS and fatigue greatly improved, and her weight normalized.

Whether or not you have any adverse symptoms from the detox-desensitization, you should be able to handle your usual daily responsibilities while on the program. You do not have to lie on a couch or relax around the house for three days. Nevertheless, I advise you to choose a two- or three-day period when your professional and personal life can be minimally scheduled. Don't just squeeze a detox program into your life. It does demand a certain amount of time and attention to be successful and meaningful.

If you don't work on weekends, I suggest you start the program on Saturday so that you are at home where you can take it easy, should the potential uncomfortable symptoms occur during the first two days. You may also feel more tired at the end of the day, so you may want to go to bed earlier than usual.

Another point to consider: You should not perform any vigorous exercise while on the detox program. Walking, stretching, and yoga are fine. Running, jogging, cycling, swimming, and vigorous tennis may tax you too severely. Because the program may make you feel more energetic than you've felt in a long while, the temptation may be there to go out and see what you can do on the jogging path or the tennis court. But please resist these temptations. Dealing with the mobilization and excretion of toxins, plus the sudden conversion from external fuel sources to partially internal ones, is more than enough to expect from your body at this time.

A detox should really be a time of peace and personal reflection. So instead of using a physical outlet for your newfound energy, why not use the energy to fortify your mental and emotional commitment to this new path of healing? Think of the detoxification as a physical embodiment of your determination to create a new direction in yourself. Use this time to read, write, contemplate, meditate, and be thankful to yourself for this opportunity to rebuild your body.

The actual modified fasting program is surprisingly simple. When you consider the shopping, cooking, eating, and cleaning up that you won't be doing, I think you'll find that the detox is actually saving you work! The specific steps include:

1. Modified fasting—no solid food.
2. Drinking diluted fruit and vegetable juices (always separately).
3. Vegetable broth to aid in cleansing.
4. Vitamin C as an internal detoxifier.
5. Powdered psyllium seed as a bulking and detoxifying agent.
6. Enemas to enhance the cleansing.

Before we get on with the specific steps of the detox-desensitization, a word of caution. NOTE: IF YOU ARE PREGNANT OR NURSING, YOU SHOULD NOT CARRY OUT THIS PROGRAM NOW. If you have any inflammatory bowel condition or any problem that raises doubts in your mind about your ability to safely carry out this detoxification program, I strongly urge you to get clearance from your personal physician before you begin. I hope that your physician is open-minded. Many conventional physicians may not have a good sense of what a modified fast or a detoxification program really is and may allow a misplaced fear of the unknown to discourage your participation. I also want to stress here that you should not discontinue any prescription medications without specific clearance from your physician.

Drink Diluted Juices to Support Your Blood Sugar

Diluted fruit and vegetable juices, which you should always drink separately at different times of the day, will be your mainstay for three days. These can be any juices that have no added sweeteners or preservatives and that you know you're not sensitive to, from previous experience. Some of the commonly allergenic fruits are pineapple, oranges, and tomatoes, so avoid those juices during your detox.

Ideally, it's nice to make the juice yourself in a juicer or buy it from a natural-foods store that stocks fresh juices. Your second choice would be bottled juice that hasn't been heated. Canned juice is your third choice. Dilute the juice with at least an equal quantity of water. For the purposes of the detox, it's best to use distilled water or a good-quality springwater during the first three days.

If you like, you may also drink carrot juice diluted with at least an equal quantity of water, or some type of juice made from green vegetables.

The purpose of the diluted juices is to prevent low blood sugar. If you have a tendency for your blood sugar to fall, you may have had an uncomfortable time when you tried fasting before. That's because your blood sugar took a nose dive on the fast, causing headaches and dizziness, weakness, maybe even nausea. With the diluted juices taken every hour or two in this program, your blood sugar should be maintained on a fairly steady keel. If you do feel dizzy, weak, or light-headed, it probably means your blood sugar is taking a little dip. That's your cue to drink some diluted juice right away. Within ten minutes, the juice should pull you out of it.

Fresh Vegetable Broth for Variety and Cleansing

If you like, you can use this vegetable broth along with your diluted juices.

Steam for five minutes:

8   small zucchini
8   celery stalks
1   pound string beans
1   medium bunch of parsley

Blend the steamed vegetables with 2 cups distilled water in a blender. Dilute to desired consistency. You can drink the broth whenever you like, hot or cold. This vegetable porridge recipe, originally recommended by Henry Bieler, M.D., is an excellent liver and body cleanser but does not support blood sugar very well.

Drinking all these liquids, you can expect to be urinating quite frequently during the detox. This is very useful. Since you'll be pulling all sorts of toxins out of your body, the increased turnover of liquid through your body will aid the flushing out of these poisons.

Psyllium Seed: An Antihunger Cleansing Agent

Go to a good natural-foods store and purchase a powdered psyllium seed product (not tablets or capsules). I ask you to do this for two important reasons. One of the major concerns of a person about to undertake a detox is, "Won't I be terribly hungry?" By using the psyllium seed according to my directions, you won't feel hungry at all.

Psyllium seed is a natural plant substance that is basically indigestible. The psyllium passes through the digestive system, like "roughage" or "fiber," and soaks up water as it goes along. It expands, produces bulk, and gives a sensation of fullness to the intestines. So when you use it, I virtually guarantee that you will not feel hungry.

The second reason for using psyllium seed is that as it passes through your digestive tract it cleans out mucus and other particulates that have been clinging to your bowel wall for months and years and have not been carried out with normal bowel movements. Most people do have small deposits clinging to their intestines, and one purpose of the detox is to get rid of them, as they interfere with digestion and assimilation.

Because of the bowel-sweeping action of the psyllium plus the nature and consistency of the psyllium itself, don't be alarmed if your bowel movements take a different shape or are abnormally colored. Keep in mind that cleansing the bowel is a crucial part of detoxification.

How to use psyllium seed: Six times a day, mix a heaping teaspoonful of psyllium seed powder with a large glass of diluted juice in a blender. Blend it for 5–10 seconds and drink it right away. (If you let it sit, it will turn into a thick gel that is not easy to get down.) Follow it with another glass of diluted fruit juice. The second glass of juice provides additional liquid for the psyllium seed to absorb.

If you don't have a blender, put the juice and psyllium powder in a jar with a tight-fitting lid and shake vigorously so that it mixes well. Then you can drink it right out of the jar. I might add that you should wash out the blender or jar right away, because the psyllium is hard to remove once it dries.

Cleanse and Detoxify with Vitamin C

While you're going through this detox, the most important supplement is vitamin C. Vitamin C is a powerful cleanser and detoxifier both for the bowels and the tissues.

When you go to your natural-food store to buy your vitamin C, try to get a calcium or sodium ascorbate powder rather than ascorbic acid. The latter is highly acidic, very tart to the taste, and potentially irritating to the stomach and intestinal lining. It tends to cause diarrhea, too. For anyone sensitive to vitamin C, I recommend a buffered form of vitamin C powder made out of sago palm.

If you have a blood pressure problem, be aware that 1 gram of sodium ascorbate powder supplies 124 mg. of sodium. This is substantial if you're trying to stick to a low-sodium diet. Personally, I have never seen sodium ascorbate raise anyone's blood pressure. However, check with your physician if you have a blood pressure question. The use of calcium ascorbate will avoid the sodium problem and is recommended for people on a low-sodium program.

Most of these powders contain between 2 and 4 grams of vitamin C per level teaspoon. I recommend that you take between 1 and 2 grams of the vitamin C crystals or powder three times a day, stirred into one of your glasses of diluted fruit juice. Stir it well so that it dissolves before you drink it. If it causes loose bowel movements or excessive gas, reduce the dose until these problems are relieved.

Consider Enemas: They Will Increase the Effectiveness of a Detox

Enemas are a useful part of any serious detoxification program. Many people can go through a detox program without daily enemas and still receive much benefit. But let me explain why your detox will be all the more effective if you include enemas.

The kidneys, skin, and lungs help out with the detoxification, but the primary responsibility lies with the liver. As toxins are released from the body tissues and processed by the liver, the liver dumps them into the bile system, which also serves as a waste removal system for the liver.

Under normal circumstances, bile is produced as an aid in digestion. But during a detox program, unless we flush this toxin-laden bile out of the digestive tract periodically, the toxins can be reabsorbed back into the portal circulation (a specific circulation between the intestines and the liver), returning to the liver when it reaches the large intestine, where bowel motility slows down. Thus, despite all your efforts to evict these poisons from your body, at the last minute they're granted a reprieve and absorbed back in.

A warm-water enema twice a day will flush out the colon and minimize the tendency for these substances to be reabsorbed through the bowel wall to go back to the liver.

Prepare a quiet and comfortable place. Put a beach towel or two on the floor and have a pillow at your head. Have some reading material at hand and/or play some peaceful music. Use the type of enema bag that can be filled to a desired quantity and that has a line with a clamp on it to stop the flow. After filling the bag with one quart of warm (not hot) water, open the clamp until you see water coming out of the end piece, indicating that all the air has been let out of the line. Reclamp the line.

Next, put some lubricating ointment or vegetable oil

on the end piece and around the anus. Fold up a large towel and place it under your left buttock so that you are lying on your back tilted to the right side. This allows gravity to work in your favor and help prevent cramping. Hang the bag about two feet above the level of your body. Hanging the bag higher than this will often cause the solution to run in too fast and produce cramping.

Insert the end piece and release the clamp. You will feel the solution start to move up your left side, then across your upper abdomen and down your right side. If you feel a cramp at any point, reach down and pinch the line for a few seconds until the cramp goes away. Then release it again.

An adult can usually handle a quart of water. It should finish going in before you feel the bowel is intolerably full. Once the solution is in, clamp the line and pull the end piece out and put it in the bathtub or sink. You may gently massage your abdomen and bowel to release any cramps. If you feel pressure to release the fluid in your rectum, raise up by lifting your buttocks off the floor so the fluid runs toward your upper body, thus relieving the pressure in the rectum. Use this trick whenever needed.

Listen to music or read, and retain the enema for five minutes while you gently massage your lower abdomen. When the time is up, sit on the toilet and gently release the fluid.

Try to do this twice a day, in the morning and at bed-time. If you can do it only once a day, morning is preferable. This is an invaluable cleansing tool, and it need be done only for three days. Remember: Enemas should be used only as a temporary, additional cleansing technique during a detox program. Never use enemas on an ongoing basis as a means of dealing with chronic constipation.

The following table will simplify the three-day first part of the detox for you. I start the table at 8 A.M. Feel free to adjust the times according to whenever you get up and go to bed.

DAYS ONE, TWO, AND THREE

8:00 A.M.	Psyllium with 10 oz. diluted fruit juice, followed by 8 oz. more diluted fruit juice
8:30 A.M.	Enema
9:00 A.M.	Vitamin C powder in diluted fruit juice
10:00 A.M.	Psyllium routine
11:00 A.M.	Vegetable broth (optional)
12:00 P.M.	Psyllium routine
1:00 P.M.	Vitamin C powder drink
2:00 P.M.	Psyllium routine
3:00 P.M.	Vegetable broth (optional)
4:00 P.M.	Psyllium routine
5:00 P.M.	Vitamin C powder drink
6:00 P.M.	Vegetable broth (optional)
7:00 P.M.	Psyllium routine
9:00 P.M.	Enema before bedtime

NOTE: You may drink additional diluted fruit juice or vegetable broth whenever you want.

Many people feel so good on the third day of the detox that they ask me if they can extend this routine a couple more days. If you find yourself feeling good and would like to add a fourth or even a fifth day to the detox program, you may. Since you're not specifically under a doctor's care, though, I would not advise going beyond five days. For those of you who are not inclined to extend the three-day routine, that's fine, too.

DAYS FOUR AND FIVE

After the initial three-day period, there comes a two-day followup and fast-breaking period. The rules are even easier than before:

1. **Psyllium seed**

   Take the psyllium seed drink only three times a day.

2. **Vitamin C drink**

   Take the vitamin C powder in diluted juices between the psyllium seed drinks three times a day, to maintain stable blood sugar levels.

3. **Discontinue the enemas.**

4. **Fruits and vegetables**

   You're now allowed to have raw fruits and vegetable salads. The fruits should be eaten *by themselves, one type at a time.* This is a precaution in case of allergies and sensitivities, which we haven't totally identified yet. If you do get any adverse reaction, add that fruit or vegetable to your list of suspect foods to be specifically tested.

   The vegetable salads can be tossed with a little fresh lemon or apple cider vinegar if you're not allergic to these. Vary the vegetables to keep the salads interesting. You may also eat vegetables that have been steamed for five minutes.

   You may eat as many meals of fruit and separate meals of vegetable salads or steamed vegetables as you desire. Always separate them by at least one hour. Chew these fruits and vegetables very well, as they have a tendency to produce gas if not digested properly.

5. **Acidophilus**

   Acidophilus (meaning acid-loving) is a strain of lactobacillus, the healthful bacteria that help turn milk into yogurt. These friendly bacteria help us in a number of ways. They control overgrowth and infection by unfriendly bacteria, and they also aid in the production of some essential B vitamins.

   Antibiotics can wipe out these friendly bacteria, as can years of eating hot and spicy foods. They are also decreased by the enemas and bowel cleansing of the

detox program—one of the very minor side effects of the
detox. But an acidophilus culture will help restore them.
Find one of the new hypoallergenic, non-milk-based
acidophilus cultures. It should contain no yeast or any
other items on your allergy list, which will be explained
shortly. Take the acidophilus in the recommended dose
starting during the two-day follow-up period. If there is no
recommended dose, take half a teaspoon of the powder
mixed in 8 ounces of water twice a day. Continue using
it on a daily basis until one bottle is used up. For more
information on acidophilus, see Chapter 6.

This five-day program should accomplish a good
detoxification for you. Keep in mind that it's not a "one
detox and you're good for life" deal. There's no way to
completely avoid harmful chemicals and pollutants.
They're in the air, water, food, medicines—literally every-
where. So I recommend that you repeat this procedure
every four months. Not only will it help keep your body
cleansed of toxic elements, it will also help you keep control over your eating habits and your weight and reaffirm
your emotional commitment to be a healthy, energetic
person taking good care of yourself.

DESENSITIZATION: HOW TO AVOID THE FOODS THAT STEAL YOUR ENERGY

The detoxification you have just carried out has helped rid
your body of toxic elements and by-products. Your body,
particularly your adrenal glands, is successfully withdrawn
from the stimulants, sugary foods, and sensitivity-provoking
foods that have been stealing your energy. Now your body is
ready to rebuild your energy supply.

But that rebuilding can be sabotaged if you reintroduce foods that provoke your food allergies and sensitivities. In order to break free of the stranglehold that these

foods have on your energy supply, you must first identify them and then eliminate them from your diet. This is the aim of the desensitization.

As the second major part of this cleansing program, the desensitization will break the hold of foods to which you've been allergic and/or addicted. It will also help identify the specific foods that are draining your energy supply.

The desensitization actually began when you started the detox. The modified fast allows a rapid desensitization. But a thorough desensitization requires a minimum of 20 days. That length of time should not present a particular hardship, because you will be eating most of the nutritious foods you want. The important part will be avoiding a list of about 20 foods on the "suspect" list. Once you've given away, stashed, or thrown away these "suspect" foods and stocked your shelves with substitutes, the rest is easy.

But first we need that list of suspect foods. One of the ways to arrive at the list is to undergo lab tests. There are many forms of lab tests for allergies. The classic type of allergy testing includes RAST tests, scratch tests, and subcutaneous injection tests. It is my experience that these tests usually identify major allergenic substances of which the patient is already aware—foods that create dramatic reactions like hives and/or swelling of the throat. These tests are also effective for identifying inhalant allergies.

There are two more useful tests to detect food sensitivity reactions. The first type is sublingual testing. After fasting, extracts of suspected foods are placed under the tongue one at a time and symptoms are noted. The patient usually gets an immediate demonstration of the food's effects, but he or she also has to live with the reaction symptoms until the tester can neutralize them. This is a time-consuming and tedious procedure, but reliable when done by a qualified tester using accepted protocols.

Another type of test is the ELISA. This is a new application of a blood test based on immunological reactions, and its technology is rapidly expanding. It is accurate, reproducible, inexpensive, and useful in determining sensitivity-provoking foods.

The above tests are good, especially for people who are not able to carry out a thorough enough desensitization and self-test. However, I strongly recommend self-testing. If you follow the directions of the detoxification and desensitization, the self-test will give you a stronger, clearer message than most medical tests. Furthermore, many people do not live close enough to a testing center, or they may simply not want to spend the money for medical testing. If you have the ELISA test performed, use your results to establish your personal list of foods to avoid.

Here's a good guideline to use in your self-testing. After years of seeing these tests done on thousands of people, I've found that these potentially allergenic foods are the culprits that show up time and time again:

LIST OF COMMON ALLERGENS

Cow's milk and all dairy products
Wheat
Corn
Soy
Peanuts
Oranges
Red meat
Yeast (brewers' and bakers')
Chocolate
White potatoes
White sugar and molasses
Bell peppers
Garlic
Eggs
Tomatoes
Black pepper
Monosodium glutamate
All food colorings and dyes
Tobacco
Beer, wine, and champagne
Coffee
Black tea

Now *add to this list* any other food you know you have
a problem with, *plus* any food you eat all the time or have
cravings for.

I'm not suggesting that you should never have a cup of
coffee or a glass of beer or wine again—or for that matter, any
of these common foods. You're just abstaining from them for
20 days in order to find out which ones are causing you
trouble and how much trouble they're causing. This is the
best way to do it.

This program is full of amazing surprises, and this is
one of the nicest: there will never be an easier time for you
to break a drug habit or food addiction than right now. Because of the unique situation in which toxins have been
flushed out of your body during your detox, toxic by-
products and residues of these substances are also quickly
eliminated. You'll find you're over any possible withdrawal
phase before you know it, with an absolute minimum of withdrawal symptoms. Furthermore, it has been my experience
that you will not have cravings while your system is cleansing.

This holds true not only for alcohol, nicotine, and caffeine (present in coffee, chocolate, black tea, and many soft
drinks), but also for over-the-counter medications and other
drugs. Many of those over-the-counter pain-killers, antihista-
mines, nasal sprays, and the like that you've prescribed for
yourself will have outlived their usefulness, anyway. More
often than not you were taking them for relief from symptoms associated with your food sensitivities—headaches,
sinus congestion, runny nose, etc. As you break the crave-
consume cycle with the detox-desensitivity program, you will
no longer need these medications. (Always check with your
physician before you discontinue the use of any prescribed
medication.)

Neither will you need alcohol, caffeine, nicotine, or any
other narcotics or stimulants. By the time you finish the total
program, you'll have energy to spare without using stimulants or other drugs. If you've ever wished that you could
wake up and start the day without cigarettes, alcohol, or
coffee, now's your big chance.

The power of this program to heal never fails to amaze

me or my patients. Harry, a 73-year-old man, complained of arthritis, fatigue, and a chronic sinus condition. I suspected that his fatigue and sinus condition were due to food sensitivity–weakened adrenal glands. But I did not think Harry's arthritis was traceable to food sensitivities. Harry began the detox-desensitization, and within five days he was free of pain and stiffness. Not only was his arthritis gone, but his energy returned and his sinus condition cleared up. Harry learned that wheat, in particular, made his joints hurt. He experienced joint pain only when he cheated on his diet and ate wheat.

Harry experienced firsthand the remarkable healing power of the detox-desensitization. You can, too.

For 20 days, starting on the day you begin your detox program, eat nothing that's on your list of suspect foods. When I say nothing, I mean exactly that—for now. During the desensitization period, there is no quantitative aspect to food sensitivities. If you are sensitive to a certain food, less than a thimbleful will set you off.

This means you'll have to start reading labels. Unless you do some detective work, you're liable to eat something you want to avoid. For example, soy and soy products are put into many prepared foods. Often, the ingredient is not even labeled as "soy." Instead, it's called "textured vegetable protein." Soybean oil is used in several salad dressings, too. Dairy products also slip into foods in a disguised manner.

You're really better off staying away from prepared foods, especially sauces, during this period. The Food and Drug Administration's GRAS ("generally regarded as safe") list allows manufacturers to use certain standard ingredients without mentioning them on the label. So you can never be sure what you're getting. You will also be better off eating most of your meals at home. In a restaurant, you're never sure what's put into the food.

I know it's going to be an adjustment for you to do without your favorite foods for these 20 days. But this is your time to say to yourself, "Courage! If this is what I need in order to feel better, then it's well worth it for me." It's only three weeks. Look on it as an investment that will bring a strong return right away: your renewed energy and health.

Since you're combining this desensitization with the detox program, by the second, third, or fourth day of the desensitization you'll see that indeed you feel much better when you're off your problem foods. If you have a lot of sensitivities, you will feel more energetic. You will have a sense of more energy, more clarity, more creativity, more life.

Besides feeling good, another bonus you can expect these first five days is to lose some weight. If you're overweight, you'll lose more than if you're normal or underweight. You'll also lose more weight, and continue losing, if you have multiple severe sensitivities. A lot of that will be water weight, because one of the things that happens when you have food sensitivities is that your cells hold abnormal amounts of water in an effort to dilute the toxic effects within the cell. When you experience rapid weight loss—5 to 10 pounds during the first few days—most of that has to be water, unneeded now that the sensitivity reactions have ended.

Over the next few weeks, you will continue to feel good and keep away the "bloat," as long as you steer clear of the foods on your list. You may continue to lose weight (if you've been overweight), because quite often the foods that make us put on weight are the foods we crave and overeat—which are also the foods we are sensitive to.

If symptoms start to recur during the desensitization period, it means either that you're not paying very close attention and taboo foods are sneaking into your diet and causing reactions; or that you're sensitive to something you haven't suspected and so have not put on your list. In this case, you're going to have to play detective again and keep a keen eye out for the culprit. If there's a specific food you seem to be craving that's part of your desensitization diet, it is a prime suspect.

Many desensitization protocols suggest staying off the suspect foods for two or three months. But if you are conscientious and do your three weeks faithfully, the shorter period will work fine. If you cheat, you'll never get the clear picture you're seeking. Do your best to do it right.

TESTING TO IDENTIFY FOODS THAT CAUSE REACTIONS

By the end of the 20-day period, you will have accomplished an adequate desensitization to the foods on your list. Of course, you don't want to stay away from all of those foods forever. But neither do you want to go right back to eating whatever you feel like and feeling lousy again. So now you're going to go through a testing procedure to identify the major affecting foods.

Every four days, choose one of the foods on your list that you would like to be able to eat again, reintroduce it into your diet, and see if you get a reaction within a few minutes to several hours after eating it. If not—and eight times out of ten that will be the case—then you may take it off your list of foods to avoid. If you do get a reaction, that food should stay on your permanent list and not be eaten. Don't worry, in all likelihood there will be only three or four foods you will need to remove from your diet.

Let's get more specific. It's Monday, and your 20 days of desensitization are up. You're ready to test the first food. You decide to test milk. So you drink a glass of milk. Ideally, have the test food by itself, with no other foods eaten for a few hours. Don't eat any other new food from the list for four more days. And don't drink any more milk the rest of Monday, Tuesday, Wednesday, or Thursday.

The purpose of delaying the tests in this way is to give you a four-day stretch for the reaction, if there is going to be one, to develop. If you were to introduce another food too soon, it would complicate matters. You'd never be sure which new food caused the reaction, since reactions can sometimes take a day to occur.

If you do get a reaction, you also want several days to clear the reaction from the body, so that you are not still feeling its effects when you test the next food.

It's very important here to use a good, "clean" test sample. Milk is obviously the best food to use to test for milk allergy. If you're testing wheat, cook up some whole-wheat cereal or wheat berries. Don't use bread to test for wheat, since bread contains a lot more than wheat: yeast, sweetener, and perhaps other grains. If you get a reaction to bread, it might be caused by any of the ingredients. Yeasts, MSG, and food colorings are available and can be individually tested.

One of two things will happen when you introduce the new food. Usually, there will be no reaction at all. If there is no reaction over the next four days, you may remove the food from your list. If you get a reaction, don't eat the food.

Rotating Your Diet

During the testing phase and after you've identified your troublesome foods, you should follow a rotation diet, even with the foods to which you are not sensitive. The diet of choice for anyone who has food sensitivities is a rotation diet, meaning that the same foods are not eaten repetitiously. The rotation of different foods minimizes the expression and creation of food sensitivities.

A rotation diet also protects against the creation of new allergies, because you've given up some favorite foods in favor of eating new and different ones. When people eat the same foods over and over, there is a tendency to develop new sensitivities, which will start you backwards into the whole sequence you just broke away from.

There may also be certain provocative foods not on the list that I gave you or that you did not consider when you made up your list of suspect foods. These will usually not be overwhelmingly obvious allergies, but they may affect you in subtle ways. By eating a rotation diet, you help to identify and eliminate sensitivities to any foods not previously identified.

Basically, the point of the rotation diet is its natural tendency to play down and eliminate sensitivity reactions. Try to eat a food only once in four days. You may eat more than one food at a time, and you can eat the same food more than once in the same day, but then try not to eat it again for four days.

Even if you don't eat a food every fourth day, which is ideal, try not to repeat it more often than every other day or every third day. In the rotation diet chart provided in the next chapter, I list foods for a five-day rotation. Thus, you have a five-day schedule from which to choose foods, even though you may be rotating your diet on a three- or four-day schedule.

With certain foods, however, you will get a reaction. It may occur within minutes to 3 hours after eating, or it may take 12 to 24 hours or longer. But if you are sensitive to the food, you will get a clear-cut reaction. If you've done your desensitization, it will be quite obvious that your body does not appreciate contact with that particular food. After three weeks of feeling so good, you'll suddenly feel as though you've "lost it." Your symptoms may be any, or possibly all, of the following: headache, irritability, joint pain, sinus congestion and draining, diarrhea, and the two classic symptoms, fatigue and the inability to think clearly. You'll know what food caused the reaction, because it was the only new element introduced to your diet within the last four days.

If you discover a food that does this to you, you should say farewell to that food, at least for the time being. Don't let that get you down. This food has been slowly draining your energy and your life force. If you continue to eat it, you can expect this fixed physiological reaction to occur every time. It will drive you back down into your addicted, exhausted state. The choice is yours. You can go back to that fatigue, or you can continue to enjoy the growth in your energy supply.

Will you have to avoid that food forever? I cannot predict that with any certainty. In my experience, as a person rebuilds and rebalances the energy system and the adrenal glands, some sensitivities will diminish. Provocative foods may then be eaten occasionally, as part of a rotating diet—certainly no more often than once every four days, and in many cases less frequently than that.

This will not be true for all people and for all foods. You may have to avoid some foods completely in order to maintain high energy. It depends on how well you rebuild yourself. There are no guarantees. You must find out on a trial-and-error basis. Every six months or so, you can try retesting the foods to which you're sensitive. If a food no longer causes a reaction, then you may reincorporate it into your diet on a rotating basis.

But please be careful. If you begin to crave the food and start eating it more frequently, it will cause you trouble.

Steps to Detox-Desensitization

I want to encourage you to plan to eventually complete the detox-desensitization exactly as described in this chapter. The vitality that will return to you will dramatically convince you that you are totally in control of your energy supply. Not only will the complete program build up your physiological energy, but your mental, emotional, and spiritual energy will also experience a boost. This part of the program can change you inside and out.

Step One: If you do not want to jump into the full detox-desensitization right away but want to ease into it, you might begin by rotating your diet on a four-day basis. You may experience some renewed vigor, or you may experience some stronger sensitivity symptoms when you reintroduce certain foods. Don't let that information throw you. Just mark it down. You don't need to make any lifelong decisions. You're only gathering information. You're learning how much power you really do have over your health and energy.

Step Two: You might think that the modified fast of the detox is the most challenging part of the program. Actually it's not. I would encourage you to try it now, for three days, with or without enemas. The choice is yours.

If the detox seems too formidable an activity to take on right now, you might instead simply try to remove from your diet as many suspect foods as you can for three weeks, while maintaining a rotation diet. This will give you additional information about what foods steal your energy.

Step Three: Until you perform the entire program as described in the chapter, you're going to get only hints and glimpses of the tremendous vitality that is your potential. For your third step, I would encourage you to approach that turning point and do the complete program. If you need additional support, use the techniques in Chapter 9, Energize Yourself from Within, to help yourself open up your mind and body to the wonderful changes that you are about to bring into being.

HIGH-ENERGY TIP

THE NIGHT-PERSON PHENOMENON

Many people claim that they are "night people," that they don't come alive until the graveyard shift. Sometimes this occurs because the individual's energy ebb and flow is sufficient but somehow reversed from normal patterns. This may account for the fact that such people have high energy at night but generally feel sluggish during the day. In my experience, however, this pattern accounts for a very small percentage of night people.

Many night people have adapted to a certain sleeping-waking cycle that allows them to have energy at night. If you normally stay up until 3 or 4 A.M. every night, naturally you will sleep later in the morning—unless you have to go to work early. In that case, the length of your natural sleep cycles is shortchanged and you feel tired during the day. Such rearranged cycles, I believe, also account for a very small percentage of night people. And even people who believe they have successfully rearranged their cycles most likely are running a borderline fatigue state.

Human beings are not naturally night people. Although some people can more or less adjust to a reversed day-night pattern, I doubt that anyone is actually physiologically designed to be more alert and energetic at night. It's easy to explain why: it's based on our ability to see. All animals that have good vision in natural light and poor vision in darkness are daytime animals. They sleep at night and begin their daily activity shortly after dawn. In the daylight, they can see, function, protect themselves, hunt, and do all the things they have to do. Because their vision is limited in the dark, they rarely operate at night. Other animals—the best example being the owl, which has excellent night vision—can be active at night because they have the ability to see in the dark.

Clearly the human being falls into the category of a daytime animal. Human beings see well in the daylight and poorly in the dark. Physiologically, we're set up to be "day people."

There is one good explanation for the night-person phenomenon. During the day, a lot of people are subject to food sensitivity–induced weak adrenal glands and symptoms affecting the brain, including the variety of problems we've mentioned: grogginess, headaches, upset stomach, indigestion, fuzzy thinking, poor memory and concentration. Such persons may eat poorly and take poor care of themselves, thus contributing to the brain fog that continues through breakfast, lunch, and dinner. Or, they may take excellent care except that they continue to eat sensitivity-provoking foods. Either way, the brain fog is extended until after their last meal of the day. However, three, four, or five hours after that last meal, they may experience a desensitization as the effects of the foods wear off. Thus, at midnight or 1 A.M. their symptoms clear up and they suddenly feel alive, awake, and clearheaded. They proceed to stay up because it's the only time of the day when they feel energetic.

What happens here is a mini-desensitization. This won't happen for everyone, because a full desensitization to a strong allergen generally takes four days. But there are some people who can desensitize faster and some foods from which people can desensitize much more rapidly. If you're a night person, the detox and desensitization will allow you to have energy all day long, whenever you need it —not just at night.

6

Up from Fatigue

"There is a natural affinity between natural foods and the human body. They share the earth as source, and the body so fed holds energies to be found in no other way."

PEARL S. BUCK

We've all heard stories about people who, in an emergency, have single-handedly lifted cars off children. In these incidents, the shock of seeing a loved one or a helpless child trapped beneath a car stimulated the person's adrenal glands and positive will to the point where he or she could summon the strength to lift the weight—before realizing that such a feat was impossible. These events, which are not always performed by healthy, physically fit people, suggest that when it comes to energy, strength, and endurance, there is really very little that is impossible for a *healthy* pair of adrenal glands in a *healthy* body.

Mind you, I'm not suggesting that if you take good enough care of your adrenals you'll be able to lift cars. But healthy adrenals will certainly supply you with enough energy to last all day without going on candy binges, guzzling coffee, or smoking cigarettes. In the event of an emergency, they'll even "supercharge" you so that you may be able to perform extraordinary feats of strength and endurance. You may not have much call for lifting cars, but you may need that extra burst of energy to get that "ton" of work off your desk.

In earlier chapters I've explained the forms of abuse that harm our adrenal and other endocrine glands and deplete our energy. I hope you've already begun the rebuilding process by identifying those areas of your diet and life that are unbalancing your endocrine glands and taxing your entire energy-supplying system.

Avoiding sensitivity-offending, stressful foods and drugs is only the beginning of your rebuilding program. Now come the foods that you're going to use to rebuild your diet, the raw materials for a new, high-energy you. First, let me say that there is no specific diet that will work for everyone. With the guidelines I will give you in this chapter, you must take the responsibility to build a diet that will work best for you, that will satisfy your needs and tastes and still keep you strong.

Before we get to the diet, it would be helpful to consider our ultimate goal, which is to eat a diet of whole foods as healthful and natural as we can make it.

Today we zip around in cars and planes and think

nothing of using space-age tools, like computers. But internally, our bodies function in pretty much the same way they did several thousand years ago, long before stoves, refrigerators, farms, and processed food. Our ancestors, anthropologists tell us, were foragers. They gathered food from the wild and subsisted primarily on roots and other vegetables, with an occasional rabbit or fish. Fruits, berries, nuts, seeds, and eggs were delicacies indulged in when they were available. Once in a while, our ancestors were lucky and fast enough to catch an animal. On these occasions they consumed the beast practically in its entirety.

Our ancestors did not have artificially fractionated foods. In their constant struggle to stave off starvation, they never would have dreamed of routinely eating just part of a food and throwing away the rest. Now that we can afford this luxury, which parts do we consume? Usually the least nutritional ones, calorie for calorie: the refined flours, sugars, and oils. Instead of eating the whole fruit, we squeeze out the juice and throw away the pulp, where all the fiber and much of the food value reside.

And the fat-marbled meat we eat today is a far cry from what our ancestors caught. The animals they ate were mostly small ones, lean and muscular because they spent most of their time running away from bigger animals. But even the bigger animals had, at most, a quarter as much fat as their modern counterparts, who are fed fat-producing growth hormones, held inactive in close quarters, and force-fed high-calorie foods.

Are you in a quandary now as to where your daily bread is to come from? Don't be discouraged. To obtain and prepare the foods for a healthy diet takes little or no extra effort. To eat healthfully is not difficult, it's just different.

You can continue to shop at your favorite supermarket. Just fill your cart with whole foods that have undergone a minimum of processing, like fresh fruits and vegetables, whole-grain products and cereals, and dried beans. Good-quality fish and naturally raised chicken are fine. Since the quality and selection of whole foods found in supermarkets varies from one area to the next, you may have to find a good natural-foods store to supply many of the items on your new diet. You may also be able to locate a local farmer or rancher who uses natural, organic methods to raise poultry, vegetables, and fruit that are free of chemical fertilizers, insecticides, fungicides, hormones, and other negative influences.

In these days of abundant food, many people eat much more than they need. One of the reasons is the addictive nature of food sensitivities. Second, most people eat empty foods, foods that contain a lot of calories but don't provide much nutrition. So after we eat them, we still get signals from the body saying, "I'm hungry, I need more food."

What the body is really saying is, "I need more nutrition." But we misinterpret this and continue to eat a lot, getting minimum nutrition and maximum calories. This makes us fat and tired, bogs down the digestion, and causes a variety of symptoms and illnesses down the line. And it will certainly take away energy and vitality.

Patients who rebuild their diets often express surprise, saying things like, "You know, I'm not eating nearly as much food as I used to, I'm holding my weight down, and I really feel pretty good. I'm really starting to enjoy some of the new dishes I'm making for myself." When you start eating useful, high-quality, nutritious food, the body becomes satisfied and you get a more honest sense of how much to eat. Your weight will adjust to the appropriate level (if your thyroid is balanced), and you will not eat what you don't need.

Such was the case with Harriet, a 33-year-old woman who came to me feeling very fatigued. Sometimes her energy would "just drop out of sight." Harriet felt tired when she went a long time without eating or skipped meals. Of course, this is a classic case of low blood sugar. As a matter of fact, Harriet's eating habits were very poor. She ate a lot of sugar. And along with that, she suffered from numerous digestive complaints and brain fog.

I oriented Harriet to a diet of complex carbohydrates to steady her blood sugar levels. I also recommended that she eat smaller meals more frequently during the day. Soon all of her complaints cleared up. Her digestive problems went away because she was no longer eating three big meals. And her fatigue slumps vanished because her blood sugar was holding steady over the course of the day, thanks to her new diet.

DIETARY RECOMMENDATIONS

Overall, I recommend a diet in which 55 percent of the calories come from complex carbohydrates, 25 percent from protein sources, and 20 percent from fats. Many foods are a mixture of carbohydrates, fats, and protein. When figuring percentages, you must decide whether a food is primarily a protein food or a carbohydrate food. Chicken, for example, may be considered a protein food, even though it also contains some fat. Beans have both protein and carbohydrates, so you may need to assign part of the calories to protein and part to carbohydrates. Don't waste energy trying to figure precisely accurate percentages—just keep close to the guidelines. Books and charts are available to help you figure this out in more detail, if you desire (see the Suggested Reading).

Complex Carbohydrates

Complex carbohydrates, such as whole grains, beans, and starchy vegetables, are digested slowly. They slowly raise blood sugar to more balanced levels and keep it there for much longer periods. Simple carbohydrates—simple sugars, such as white sugar, molasses, ripe fruits, and honey—are digested rapidly, so eating too much of them will boost blood sugar to disturbing levels and upset our energy supply by upsetting the body's blood sugar balance mechanism.

Some parts of complex carbohydrates are not digested at all: the fibrous parts, the "roughage." They may not provide calories or food value, but they can lower blood levels of cholesterol, help prevent bowel cancer by absorbing toxins and preventing constipation, and help satisfy hunger by giving a full feeling. The bran portion of grains supplies the most fiber. Wheat bran, oat bran, corn bran, and rice bran are the most commonly available forms of cereal fiber. The fibrous parts of fruits and vegetables also supply fiber.

Until relatively recently, complex carbohydrates were the principal source of nutrition and energy for many humans. You can—and many people in the world do—live a healthy and vigorous life eating a diet in which more than 75 percent of the calories come from complex carbohydrates. In countries where use of these foods far outweighs the use of dietary fats, animal proteins, and sugars, there is a wealth of evidence that they help prevent degenerative diseases such as cardiovascular disease, arthritis, diabetes, cancer, and premature aging.

Protein

The quality of protein you select makes a big difference to your general health and to the health of your adrenal glands. My guidelines for protein foods are as follows:

1. Avoid red meat. It is too high in saturated fat and LDL (low-density lipoprotein) cholesterol, the type of cholesterol that is deposited on the walls of blood vessels and is known to increase the risk of heart attack and stroke. Red meat is also harder to digest and often produces gas and a bloated feeling. Red meat is also prone to putrefaction in the bowel, which has been linked to constipation and other more serious bowel diseases. Some of the by-products of this putrefaction are cancer-causing, so it's not surprising that there are statistical links between bowel cancer and diets high in red meat.

2. For your animal protein needs, use fish and fowl. Cold-water fish are especially good because they contain higher levels of certain oils believed to be protective against cardiovascular disease (see the following section on fats and oils). Cold-water fish

are those that have a fishy taste, such as cod, mackerel, salmon, and sardines.

3. To cook animal protein, bake, broil, or steam—do not fry or deep-fry. Be sure not to overcook protein, as its nutritional value is then decreased.

4. If you are a vegetarian or are considering becoming one, please use good judgment. Don't become a macaroni-and-cheese vegetarian. Don't trade your fish and chicken for substandard nutrition. Use grain-bean combinations to make sure you get enough high-quality protein to maintain good health and energy.

In my experience, the average person of European descent feels better and maintains better health when discreet amounts of animal protein are included in the diet. Nevertheless, you *can* get enough protein from vegetarian foods. A grain-bean combination is the main protein source for half the people in the world. In Central and South America, it's corn and beans. In the Orient, it's rice and soybeans or soybean products. A grain-bean combination has a lot of nutritional advantages. It provides more usable protein than either grains or beans alone and provides a good amount of fiber. And since it contains complex carbohydrates, it stabilizes your blood sugar level and prevents degenerative disease.

Be mindful of your food sensitivities when combining grains and beans. Most beans are only occasionally a problem, except for soybeans, which are potentially allergenic. A good vegetarian cookbook will give you dozens of ways to serve grain-bean combinations. (I recommend my favorites in the Suggested Reading list for this chapter.) Just substitute for or leave out those ingredients to which you are sensitive. My favorite way to prepare grain and beans is very simple: cook them up slowly, mix them together, serve them with steamed vegetables on top, and season with any herb (or vegetable sauce, like tomato) that's not sensitizing for you. Remember that carbohydrate digestion begins in the mouth.

So to ensure complete digestion and minimize gas, chew
your foods well and mix them thoroughly with saliva before
swallowing.

Fats and Oils

Excess dietary fat ultimately deprives us of energy and
life. Continual high levels of fat in the bloodstream affect the
red blood cells, which are responsible for carrying oxygen to
other cells. The affected cells stick together like stacks of wet
dinner plates. This not only diminishes their effectiveness in
carrying oxygen to the tissues, but also hinders their very
passage through the small blood vessels. This continual low-
grade oxygen starvation of the tissues could be a primary
cause for many degenerative diseases.

Even if you reduce your fat and oil consumption to no
more than 20 percent of your total caloric intake, the kind of
fat or oil you're eating is just as important as how much you
eat. Most of these oils can be dangerous to your health if
eaten in excess. But some fats and oils are decidedly more
dangerous than others, while some are necessary and beneficial. To begin with, man-made, refined fats and oils are more
dangerous than natural fats. Hydrogenated vegetable short-
ening, margarine, and most foods that contain them, such as
chips and many baked goods, are chemically foreign to the
body. Saturated fat is usually animal fat, but it also includes
certain vegetable oils such as coconut and palm. Most vegetable oils are polyunsaturated, unless they are artificially saturated through hydrogenation. Some, like olive oil, are mo-
nounsaturated.

My guidelines for fats and oils are as follows:

1. Avoid butter and margarine. Butter contains
    saturated fat and LDL cholesterol. Margarine
    contains lots of artificial chemicals and additives and
    should be totally avoided. The oils in margarine are
    very unstable and, when heated, can deteriorate into
    cancer-causing chemicals. Butter and margarine can
    be replaced by small amounts of vegetable oil for

cooking or baking, by pure fruit purees on bread or toast, or eliminated altogether.

2. Avoid fried foods. Heating to high temperatures not only reduces the value of the protein in the food but alters chemical bonds in the fats and oils. Unsaturated oils can become saturated, and unstable oils and fats can form cancer-causing compounds.

3. Eggs are high in cholesterol. But studies show conflicting results as to whether eating eggs actually raises blood levels of cholesterol. Until there is more definitive evidence, I suggest that if you have a cholesterol problem, you should avoid eggs entirely. If you do not have a cholesterol problem, you may eat two eggs two or three times a week. Do not eat hard-boiled or fried eggs. Eat them poached or soft-boiled, with the yolk still liquid. When the yolk is hard, it's a sign the protein has been overheated and is therefore less valuable and harder to digest. Yolks can be eaten raw, but the white of the egg should never be eaten raw. Raw egg white contains a chemical substance that inactivates a vital B vitamin, biotin.

4. I recommend that you avoid cheese. It is a man-made milk product and, as such, is an incomplete food. It is highly concentrated in fat and tends to be constipating and fattening.

5. Some oils, such as safflower, peanut, and sunflower, are sold in health food stores as natural foods. Nevertheless, they are actually concentrated to an unnaturally high degree. In nature, oils do not occur apart from the food source from which they are derived. Our ancestors would never have squeezed the oil out of something and thrown away the rest.

Some oils, however, are beneficial when consumed in small quantities. The essential fatty acids, linoleic and arachidonic, are required for production of hormones, cell membrane structure, and the health of the immune system. Without small quantities of these oils in our diet (from vegetable oil sources, such as linseed, safflower, and sesame) we could not survive. EPA and DHA oils (eicosapentaenoic acid and docosahexaenoic acid) are oils derived primarily from cold-water fish (cod, salmon, mackerel, sardines). They have been found to help protect against cardiovascular disease and, in high enough doses, are anti-inflammatory. Eating the flesh of these fish supplies small amounts of these oils.

All oils must be kept tightly capped and in a dark, cool place to avoid becoming rancid. Because olive and sesame oils are more stable to heat and are thus much less likely to become rancid, they are the safest and most practical to use. But remember, oils should be used sparingly, as in home-made salad dressings. Remember also that oil-based salad dressings can be high in calories and should be used as little as possible if you are trying to control your weight.

Stop Sugaring and Salting Up

I believe you should not only cut white sugar out of your diet totally but also reduce your consumption of all other sugars as well. That includes natural sugars found in fruit juice, honey, maple syrup, fructose, dates, figs, other dried fruits, and corn syrup. Avoid all junk food, not only because of the high sugar content but also because of the high fat and salt content.

In Chapter 4, we've already talked about the effect all forms of concentrated sugars have on blood sugar and energy supplies. Though you may get a quick jolt of energy, you're soon down on your knees energy-wise begging for more sugar.

You don't have to avoid fruit juice entirely. My general rule is to dilute all fruit juice and carrot juice with equal parts of water or seltzer. You can dilute carrot juice with green vegetable juice if you prefer. Better yet, eat the whole food.

But even when you eat the whole fruit, you're better off eating it before it gets overripe. As fruit ripens, the complex sugars are broken down into simple sugars, so the stress on the body increases.

Cutting back on salt (sodium) often has beneficial

effects on high blood pressure, migraine headaches, and vascular disease. I recommend that you wean yourself from salt as soon as you can. If you are eating fresh, green vegetables consistently, you will get plenty of sodium. Judicious use of a little sea salt now and then is all right, unless you have high blood pressure or must follow a low-sodium diet.

High-sodium foods include salted or smoked meats and fish, cold cuts, processed or strong cheeses, salted snacks, olives, pickles, most commercial sauerkraut, processed soups and bouillon. Baking powder and baking soda contain almost as much sodium as table salt.

To Drink or Not to Drink

Break the habit of drinking beverages with your meals. Drink between meals instead. A small amount of liquid, such as a small teacup's worth, is all right with a meal. But any more than that will dilute your digestive enzymes, cause indigestion and incomplete assimilation and utilization of the nutrients. Drink your liquids one half hour before or two hours after your meals.

Water is a necessary nutrient. The amount you need depends on the weather and your level of activity. It's preferable to drink between one and two quarts of liquid daily, unless your physician has advised you otherwise. If your municipality doesn't supply good-quality, pure water, and if you don't have your own well, then you should drink bottled springwater, or filter your tap water through a solid block charcoal filter (see *Nontoxic and Natural,* by Debra Lynn Dadd, for a description of water and water filtration systems).

EATING MEAL BY MEAL

The above guidelines are only general. Now we're going to get more specific, on a meal-by-meal basis. Please be mindful of your food sensitivities when following these recommendations. Do not use any food that provokes a food sensitivity reaction, and remember to try to rotate your foods as described in Chapter 5. Choose your energy-building foods from Table 6-1 and use Table 6-2 to help you rotate your foods.

The kinds of foods we eat at breakfast, lunch, and dinner are largely determined by culture. If you feel best when you have a salad for breakfast, broiled fish for lunch, and a whole-grain pancake for supper, go right ahead. Don't feel limited by past patterns that may have contributed to your fatigue. Your new energy-building diet is an adventure.

Breakfast is, just as we were told in grade school health class, the most important meal of the day. It sets things in motion metabolically for providing daylong energy. You should never skip breakfast, especially when you're rebuilding your adrenal glands or if you have a tendency toward low blood sugar. Breakfast never makes you fat. You use its calories to accomplish your daily tasks.

A bowl of hot, whole-grain cereal is a nutritious breakfast choice. Mixing seeds such as sunflower seeds, pumpkin seeds, sesame seeds, and chia seeds with your whole grains is a nutritionally useful practice, because the seeds help balance out the incomplete amino acid profile of the grains, providing good-quality protein from a nonanimal source. The types of seeds, like the grains, should be rotated in your diet.

For a light lunch or dinner I would suggest a salad of raw vegetables. Don't think of a salad as just lettuce and tomatoes. Any raw vegetable can be used. Add some fish or chicken to the salad for basic protein. Or have a grain-bean combination with the salad. Commercial salad dressings can be a problem. Most are loaded with fat and sugar. So it's best to make your own with lemon, vinegar, or a non-oil base like tomato juice. For variety, a little olive oil once in a while is all right.

Other examples of grain-bean combinations include soups and stews like barley-and-lentil soup. You can also cook up the grains and beans, then mash them up with seasoning and other ingredients and bake for 20 minutes at 350° to make "vegie" burgers. Also remember that pasta and wheatless pasta (for wheat-sensitive people) can constitute the grain portion of the combination. The grain and beans need not be in the same dish, but are best eaten at the same meal. You could include beans in your salad with the pasta meal or simply put them in the sauce you prepare.

Several possibilities exist, and the cookbooks recommended will assist you. Remember to avoid sensitizing foods and also to be creative and use your imagination. You'll find your own tasty way to prepare these dishes, and you will enjoy the energy they consistently create for you.

For a main meal, choose cold water deep-ocean fish or organically raised fowl (skin removed), which are comparatively pollutant-free, are not typically sensitivity-offending, and are lower in saturated fat than beef, pork, or lamb. Avoid the routine eating of shellfish: not only are they high in cholesterol, but since they are shore scavengers, they tend to pick up mercury and other pollutants. It has been recognized for many years that mercury compounds contaminate our water and soil and that large predatory fish tend to accumulate large amounts of them. The dangers of mercury toxicity include many neurologic disorders and retardation of children. Swordfish is the worst offender and should be avoided by pregnant and nursing mothers. *Limit* tuna, halibut, red snapper, perch, and pike. Lake fish in the eastern United States have been found to be contaminated with various organic compounds like PCB, dioxin, DDT, and chlordane. Farm-raised fish, usually trout and catfish, are generally safe to eat, but *avoid overeating* trout, carp, catfish, chub, striped bass, whitefish, and American eel that are caught in eastern lakes.

Typically, supper is a very high calorie meal, eaten too late in the day for the calories to be digested, absorbed, and burnt up before bedtime. This is the kind of meal that makes you fat and decreases your energy supply. So my recommendation is to avoid rich, high caloric-density foods at your final meal of the day. If you go to bed before one-third to one-half of your entire day's caloric input is even digested, let alone burned, those calories are stored before they've had a chance to be used. Not only will weight loss then be difficult, but you'll feel tired and sluggish when you wake up in the morning. Eat a light dinner early in the evening and you'll wake up feeling light and alive.

Remember that too much animal protein is a burden on the metabolism and digestion. So if you do choose some form of animal protein, stick to small portions of 6 ounces or less.

Snacks are optional unless you experience low blood sugar. In that case, I recommend a snack in the late morning and then again in mid-to-late afternoon. Good snack foods include fresh fruit, raw vegetables, diluted juice, our diluted vitamin C juice drink described in the previous chapter, and popcorn (no butter or salt!). A good snack is one that works for you, rather than against you, to supply energy and reduce your appetite. I generally don't recommend nuts, as they are difficult to digest and are fattening. Use them sparingly.

TABLE 6-1
ENERGY-BUILDING FOODS

Complex Carbohydrates

whole grains (wheat, rice, corn and cornmeal, oats and oatmeal, rye, buckwheat, triticale, barley, millet)
starchy vegetables (potatoes, sweet potatoes, yams, winter squash)
whole-grain pasta, bread, crackers, cereal
all beans

Vegetables

artichokes, asparagus, bean sprouts, beets, bok choy, broccoli, Brussels sprouts, cabbage, carrots, cauliflower, chard, chicory, cilantro, celery, cucumbers, endive, escarole, eggplant, green pepper, greens (collard, dandelion, kale, mustard, and turnip), jicama, lettuce, mint, mushrooms, okra, onions, parsley, radishes, rhubarb, rutabaga, spinach, string beans, summer squash, tomatoes, turnips, watercress, zucchini

Fruits

apples, apricots, bananas, berries (blackberry, blue-
berry, raspberry, strawberry), cherries, cranberries,
grapefruit, grapes, mangoes, melons (cantaloupe, hon-
eydew, watermelon), kiwi, nectarines, oranges, pa-
payas, peaches, pears, persimmons, pineapple, plums,
prunes, tangerines.

Protein

fresh fish, chicken, turkey, eggs, occasional veal, milk
products

Sweeteners

small amounts of honey and pure maple syrup, fruit
juice and juice concentrates for baking

FROM NOW ON—EAT TO LIVE

From now on, you're expecting something more than a
short-range gourmet experience from your food. You're expecting health and vitality. You're going to eat to live, not live
to eat.

This doesn't mean you're not going to enjoy your
meals. Enjoyment of our food is part of the benefit the food
gives us. As your new, conscientious diet starts to pay divi-
dends and you begin to look and feel younger and more
vigorous, you will find yourself enjoying your meals more
than you ever did before. You'll be able to eat with the knowledge that your food is building health and supplying you with
energy.

As you continue with your high-energy diet, you'll
notice your tastes change. As you commit to the healing path
and gain accurate information and intuition about what foods
do and don't work for you, as you begin to purify and rebuild
your body, your interest in foods that aren't good for you will

ROTATION DIET

FOOD GROUP	DAY 1	DAY 2	DAY 3	DAY 4	DAY 5
Grains	Buckwheat	Barley	Millet	Oats	Brown Rice
Seeds	Pumpkin	Sesame	Sunflower	Chia	
Nuts	Filberts	Cashews	Brazil	Walnuts	Pecans
Melons	Canteloupe	Water-melon	Crenshaw	Honeydew	
Fruits	Banana Papaya Boysen-berry	Apple Apricot Coconut	Tangerine Blackberry Nectarine Pineapple Plum/Prune	Cranberry Grapefruit Strawberry Gooseberry Lemon	Peach Cherry Blueberry Lime
Vegetables	Artichoke Spinach Celery Parsnip Avocado Chives Sweet Potato (Yellow)	Romaine Lettuce Carrot Endive Asparagus Leek Watercress	Broccoli Cabbage Rhubarb Water Chestnut Collard Swiss Chard Olive	Butterhead Lettuce Cauliflower Turnip Mushroom Winter Squash Eggplant	Red Leaf Lettuce Kale Radish Cucumber Summer Squash Yam
Legumes	Alfalfa Peas	Kidney Beans Lentils	Lima Beans Split Peas	Mung Beans Black-eyed Peas	String Beans Garbanzos
Fowl	Chicken	Duck	Turkey	Wild Game Fowl	Cornish Game Hen
Fish	Bass Salmon Sardine	Catfish Red Snapper Smelt/Shark	Flounder Perch Sole	Halibut Mullet Cod	Herring Mackerel Trout
Seasonings	Turmeric Nutmeg Thyme	Ginger Clove	Caraway Paprika Oregano	Cinnamon Sage	Mustard Vanilla Curry

Take 15–20 minutes at the beginning of every week and use this chart to help you personally plan your rotation of foods and menu ideas. This will simplify both shopping and food preparation. It will also help you incorporate all my suggestions without any mystery.

drop away. In a short while, you'll wonder how you could have liked them so much in the first place.

Likewise, your tastes will change in regard to certain foods you don't really like now but which you know are good for you. You'll acquire a taste for them and genuinely come to enjoy them. This will be a real turning point. Encourage yourself to grow into new, healthy food preferences. As you learn more and grow more, your energy supply will increase, too.

Steps to Rebuilding Your Energy Supply with the Right Foods

It may seem as though the changes you are about to make are the most formidable you've ever made in your life. They're really not. When a new type of restaurant opens in town, serving a variety of ethnic foods that you've never tasted before, you don't let that stop you, do you? Think of your new high-energy diet that way. Your new menu will give you more health and vitality than you've ever experienced.

Step One: Make an inventory of your diet. Keep a food diary listing everything you eat for a week or two. This will help you become conscious of what foods you're putting in your mouth. If you've already gone through the detox and desensitization, you're already pretty aware. If not, this step will provide you with some helpful information.

As you take this inventory, you may become aware of which foods are healthful and which are draining your energy. There may, of course, be some mysteries yet to be uncovered by the detox and desensitization, some common foods to which you're sensitive. Draw a red line through those foods that you know are draining your energy. Any food you crave or eat frequently should be eliminated. You may not banish them completely from your diet quite yet, but at least you'll be honest with yourself. Resolve to do all you can to rid your diet of junk food.

Step Two: Carry your resolve a bit further by adding to your diet foods that this chapter identifies as healthful and energy-building. You will find there will be several new foods you enjoy. You will also realize that you'll need to eliminate other foods along the way to make room for new favorites. Eliminate the unhealthful foods, the ones you drew a red line through in the previous step.

Step Three: As you carry out my suggestions, you'll notice that you become more and more aware of the health value of what you're eating. You will notice which foods make you feel tired and which ones help you feel energetic, healthy, and alive. The concept of "discipline" will disappear as you automatically drift away from eating habits that don't work and move toward those that do. This is exactly what you want to happen. Once you reach this point of awareness, all it takes is your decision to go all the way and follow your own knowledge to its logical conclusion. If you want true high energy, you will go for it.

HIGH-ENERGY TIP

THE AIKIDO PRINCIPLE—MAKING USE OF THE ENERGY AROUND YOU

The energy around you given off by people and activities can nourish you and give you energy just as surely as food. Aikido is a technique built on this principle. Aikido is one of the more passive martial arts, which teaches a system of defense based on turning the energy you're being attacked with against the attacker. You use the other person's energy. This is a metaphor that applies to all life situations, not only those involving self-defense. You can use the energy of life, the energy around you.

To accomplish this, you must understand that all the energy you manifest doesn't have to be taken out of your own body. My purpose in this book is not to create a huge warehouse of physical energy somewhere inside you that you must draw on throughout the day. You can use the energy inherent in particular situations for your own mental and physical energy. Part of the art of living and of being a peaceful and vital person is to learn how to channel these useful and positive energies around you instead of always using your own personal physical energy.

The simplest example of this phenomenon is performing before an audience. Actors, actresses, and athletes will tell you that performing in front of an audience is more energizing than performing in front of a camera without an audience. The audience itself supplies energy. They are paying attention and focusing energy on the performer, who can feel it and channel it through himself and circulate it back to the audience. The performer becomes a conduit for enormous amounts of energy—energy from hundreds, perhaps thousands, of people.

An athlete playing before a crowd—even a small crowd, such as a high school athlete playing before a few dozen students and some parents—receives energy. All of those people are focusing on the athlete, giving him energy, cheering him. Even if they're booing, it's energy directed toward the athlete, who can take it and use it positively.

You can learn to use the energy around you, even if it's coming from only one person. When people give you compliments, they are giving you energy. Don't discount a compliment, take it in, feel it, and use it. If you are in an important business meeting, feel the energy around you, and instead of resisting it, relax with a few deep breaths and again decide to feel it and use it. Circulate the energy that's in the life forces around you.

7

Supplementing Your Diet

"No nutrient by itself should be expected to prevent or cure any disease. Nutrients always work cooperatively in metabolism as a team."

HEFFLEY AND WILLIAMS

Whenever I talk to my patients about supplements, these basic questions often come up in one form or another: "Why do I need to take supplements? Why can't I get all the vitamins and minerals that I need from my food? Where were the supplement stores 100 and 500 years ago? How did people survive so well for so long without supplements?"

These are all reasonable questions. The answer is twofold: insult and agriculture. No matter what kind of life you lead today, you are exposed to potentially harmful chemicals and oxidants in the air, food, and water. In our emotional as well as our physical lives, we are subject to a lot more stress. All of these stresses are parlayed into the body through the adrenal glands, the organs principally responsible for handling our response to stress, which includes maintaining our energy supply. So not only does the stress of modern living use up more of the metabolic ingredients necessary for life, it also uses up more of our fundamental life-giving energy.

That's insult. Now for agriculture: These metabolic ingredients, or nutrients, that we need to live a healthy, vigorous life—and ultimately, to live at all—are not as available in the food chain as they were 100 or 500 years ago. Today, "fresh" fruits and vegetables are grown with the aid of chemicals designed to develop large, heavy—but not necessarily nutritious—produce, picked before they're ripe and stored for many days before they're shipped by train or truck. Then they lie in the supermarket warehouses for a few days before they find their way to the market shelves.

Many studies have been done on the content and availability of nutrients from "fresh" foods taken off supermarket shelves. It's hard to believe, but oranges have been found with hardly a trace of vitamin C, and carrots and squash with little vitamin A.

RDAs—FACT, FALLACY, OR FANCY?

The requirements for basic nutrients are the RDAs—so-called "recommended daily allowances." The RDAs are absurdly low. The original intention of establishing RDAs was to prevent deficiency diseases like scurvy, pellagra, and beriberi. What's interesting is that doctors continually say there is adequate nutrition in the normal American diet, meaning that the RDA doses of vitamins and minerals are present. The RDAs are so low that many times this will be true.

On the other hand, many people are not even getting the RDAs. I have done computerized analyses of the diets of several thousand people who had listed all the foods they typically eat in a month I ran the diets through a computer and analyzed the amounts of carbohydrates, protein, fat, vitamins, and minerals. For many of the B vitamins in particular, for vitamin D, vitamin E, sometimes vitamin C, and for many minerals, I actually found many people getting less than the RDA doses for these vital nutrients.

In addition, scientists such as Roger Williams, Ph.D., have demonstrated that people have "biochemical individuality." That is, we all have different needs, and the true daily requirement for one person might be 20 to 100 times what it is for another. Furthermore, most of us who want high energy want more than the base level of health that RDAs were set up to maintain. Therefore, we need to take supplements not only to make sure we're getting the RDAs, but to maximize our potential for energy.

SUPPLEMENTS HELP PEOPLE GET BETTER

We know that certain factors, such as stress, tend to create breakdown in the body. We also know that certain dietary supplements, such as vitamins E and C, can help heal many diseases. Not only have people said they feel better and had their diseases get better, but physicians and trained researchers have observed these results. We now know that nutrients help us to resist the factors that create breakdown and eventually cause diseases. Supplements do exactly what their name implies: they shore up and add extra strength to our physiological resistance to all stress.

As valuable as supplements are, they cannot do the job alone. Do not make the mistake of believing that if you take enough supplements you will be able to ignore food sensitivities, eat all the junk food you want, neglect exercise, and generally take poor care of yourself. Supplements are a vital part of any overall health plan. In some cases they may be key factors in rebuilding energy. But they are seldom the only factor.

Nicholas, a 37-year-old man, complained of fatigue and digestive problems. His diet was good but he was not taking any supplements. I instructed him in the detox and desensitization, which created a slight improvement. Then I prescribed digestive enzymes, vitamins, and minerals. He soon experienced relief from his symptoms. His energy returned and his digestive problems ceased. Nicholas reported some months later that the supplements definitely made the difference. He had more energy when he took them and felt tired when he skipped them for a couple of days.

TO HEAL AND MAINTAIN: VITAMINS FOR VITALITY

In a table later in this chapter, recommended dosages of many vitamins are listed in two forms: healing or therapeutic dose, and maintenance dose. These dosages will, in most cases, be high enough to account for biochemical individuality; however, some people may get by with less and some may require more. At some point your "therapy" will be complete. Nevertheless, you will want to maintain a high level of nutritional support to keep your health on a positive, forward-moving track.

The crossover point from healing to maintenance comes when you have a sense that you've pulled out of the acute phase of your fatigue. When you're nearing the crossover point, you'll feel like you're rebuilding. You'll feel good again, stabilized in the rebuilding process. Your cravings for sensitivity-provoking foods will be disappearing. This is the point at which you can start thinking about crossing over to the maintenance dose.

The detoxification and desensitization described in chapter 5 will get you to a point of accurate communication with your own body. You will find it easier and easier to get clear and accurate feedback, so please do that part of the program before getting into supplementation.

I strongly urge you to obtain the guidance of a medical doctor who is trained in the use of nutritional supplements. Such medical doctors are rare, but they do exist.

Before I recommend specific supplements and doses, I want to give you a few general rules.

Unless otherwise stated, all of the following supplements should be taken with meals. Because these are all concentrated food nutrients, you want them to be subject to the same digestive processes that food itself is, including the action of enzymes and other digestive substances. When you merely swallow your supplements with juice or water between meals, the entire digestive process does not take place. The body senses only that it's getting a drink, not a meal. Not only is the level of assimilation higher when you take supplements with a meal, but they are far less likely to cause nausea or some other digestive upset than when taken on an empty stomach.

The supplements are to be taken, whenever possible, in two divided doses. For example, if I recommend 100 mg., take 50 mg. twice a day. You may take the supplements in individual tablets or capsules or as part of a multivitamin-mineral supplement.

In general, if you are sensitive to ingredients in supplements—and many people are, the best example being the yeast in B vitamins—be sure you're taking hypoallergenic supplements. Most natural sources of B vitamins made with brewers' yeast have considerable allergenic potential. There are rice-based B vitamin supplements, but avoid them if you have a rice sensitivity. I would prefer that you use natural supplements, but I use synthetic B vitamins with my allergic patients more and more, in order to avoid hidden sensitivities.

Also, it's impossible to get really high doses of B vitamins in anything but synthetic form. Even though brewers' yeast, rice bran, and malt extract are the substances in which the highest concentrations of B vitamins are known to occur naturally, that's on a relative basis. On an absolute basis, the amounts in those substances are quite small.

As is true of most things in life, you get what you pay for. Buy reputable brands of supplements in a form to which you are not sensitive. If you have trouble locating them, or are unsure about quality or allergenic potential, write to me and I will tell you how to obtain good supplements.

Whole books have been written about individual vitamins and minerals. In this section I am trying only to present the vitamins and minerals in simple fashion as they relate to energy. For more information, there are many excellent books available (see the Suggested Reading list).

At the end of the sections in this chapter on vitamins and minerals, you'll find tables listing all the recommended dosages.

Vitamin A Besides being a potent antioxidant, vitamin A is required for the maintenance of the adrenal glands and the integrity of the skin, hair, and immune system. It was once thought that the first sign of vitamin A deficiency was loss of night vision, since vitamin A is also vital to the healthy structure and function of the eyes. However, recent research has demonstrated that the first sign that the body is not getting enough vitamin A may actually be anemia, or a reduction in the ability of the blood cells to carry life- and energy-giving oxygen to the cells of the body. So an inadequate intake of this vitamin can not only result in reduced protection against disease, but also increased fatigue.[2]

There are two forms of vitamin A. Fish-liver oil is preformed vitamin A. Beta-carotene is a provitamin, a substance that our body converts to the active form of the vitamin. Beta-carotene has recently been shown to help protect the lungs from oxidant damage (smog is high in oxidants), cancer and other diseases, and also to provide strong support for the immune system. It has less toxic potential than the preformed vitamin because the body must first convert it to the active form and also because it is less effectively assimilated. It does not go directly to the liver as does fish-liver oil. Beta-carotene first gets a pass or two through the body tissues.

I recommend taking both forms of vitamin A to ensure that all aspects of vitamin A needs are handled. The dose of vitamin A I recommend is 10,000–20,000 IU (international units) a day. A word about vitamin A toxicity: The toxicity of vitamin A is generally greatly overstated. There will be no problem with beta-carotene in doses of 10,000–20,000 IU per day. If you are taking both beta-carotene and fish-liver oil, I recommend taking 10,000 units of each. This dose is appropriate both for the therapeutic and maintenance period.

Vitamin D This vitamin regulates calcium absorption and utilization. Supplementation is not necessary in a sunny climate. If you live in a place where you are regularly exposed to the sun, you probably don't need a vitamin D supplement. If you do not live in a sunny climate, or if you do not spend much time outside during the day, a supplement of 400 IU will suffice. Generally, about 400 IU of D comes along with each 10,000 IU of vitamin A in fish-liver oil.

Vitamin E Vitamin E is another potent antioxidant. Inadequate intake results in shortened lifetimes for the red blood cells, diminished function of the pituitary and thyroid glands, and slow degeneration of the brain, spinal cord, endocrine glands, and muscles. In light of these biochemical facts, vitamin E is extremely important to our energy supply. I recommend that 400–800 IU of vitamin E be taken each day for the therapeutic phase and 400 IU daily for the maintenance dose.

Mary is a 38-year-old woman who complained of fatigue and fibrocystic breasts. I found that she had very tired adrenal glands. She was drinking a lot of coffee to help her get through the day and handle her responsibilities. She was not taking any vitamin supplements. We now know that drinking coffee is linked to fibrocystic breasts and that lack of adequate vitamin E can also encourage this condition. I instructed Mary to improve her diet, stop drinking coffee, and take daily supplements of 800 IU of vitamin E, along with other basic supplements. Mary's breasts are now normal and her fatigue is greatly improved.

Vitamin K

Vitamin K is a factor in blood that promotes coagulation and clotting. It is not available without a prescription. If you have a combination of fatigue and bruising easily, you may need extra vitamin K. Bruising easily is usually attributed to a deficiency of vitamin C and zinc. But if you're taking vitamin C and zinc and still bruising easily, you probably need more vitamin K—which you can get by eating extra broccoli, turnip greens, cabbage, and spinach. You can also get a prescription from your doctor for a small dose of vitamin K. Most people are not deficient in vitamin K, and fatigue as a result of vitamin K deficiency is not a common problem.

The B Vitamins This group of vitamins has traditionally been associated with our energy supply. Standard textbooks on nutrition always state as one of the prime functions of the B complex: "the release of energy from food." The B vitamins are intricately involved in the digestive and metabolic processes. For our purposes in this book, it is enough to know that these processes include carbohydrate, protein, and fat metabolism. Most of the biochemical reactions that make up the energy-producing system require one or more B vitamins somewhere along the way. Several, if not all, of the B vitamins are also vital to the function of the adrenal glands and the liver, two of the most important organs in the energy-producing system.

And because the B vitamins are water soluble, they are not stored for very long in the body. Therefore, if you're undersupplied with B vitamins, chances are that you'll run out of energy before other symptoms start to appear.

Let's take a look at specific B vitamins and, in brief, their role in producing energy.

B_1 *(thiamine):* Thiamine can be likened to the "spark plug" of the body's biochemical engine. As a coenzyme it plays a vital role in the burning of carbohydrates for energy. Our bodies run best using unrefined fuel, like complex carbohydrates, which the body refines into blood sugar before using. When we use highly refined fuels, like white sugar, the body's energy flame temporarily burns with a higher intensity. The flame soon burns out—but before it does, it uses up an abnormally high amount of the thiamine available for energy. So the use of white sugar and other highly refined carbohydrates can rapidly produce a deficiency of this vitamin.

B_2 *(riboflavin):* Riboflavin is required for cellular respiration. Without adequate riboflavin, the cells simply don't metabolize well. Obviously, cells that are subtly suffocating are not going to be able to produce much energy.

B_3 *(niacin):* Niacin is also vital to cellular respiration. Actually, niacin is required for the cell's utilization of all major nutrients. A niacin deficiency often results in deterioration of gastrointestinal tract function. This deterioration often produces increased tendencies to develop food allergies. Because of the breakdown in the integrity of the intestinal wall, certain protein macromolecules gain access to the body and provoke a sensitivity reaction. Normally, these proteins would not gain such ready access and the sensitivity would not develop quite so easily.

Niacin is not toxic to generally healthy people. However, high doses sometimes cause a so-called niacin flush: 15 to 30 minutes after taking it, you may get red and hot and itch a little bit. This is from histamine being released by the niacin. It doesn't hurt you but it does feel a little strange. You can gradually build up your niacin dose. Add 25 mg. each week until you're up to 100 mg. with only mild flushing. Continue that through the maintenance period. If the niacin flush is a problem for you, you can avoid it totally by using niacinamide, which is actually the form of the vitamin that exists in the human body and in other animals. Niacin, or nicotinic acid, is the vegetable form of the vitamin.

Pantothenic acid: Of all the B vitamins, pantothenic acid is most directly associated with adrenal structure and function. This has been vividly demonstrated in a classic experiment involving five groups of rats swimming in cold water. The first group of rats was deprived of pantothenic acid. They were able to swim for only 16 minutes. The second group of rats each received the rat's "RDA" for pantothenic acid—and stayed afloat for 29 minutes. The third group of rats was fed "excess" amounts of the vitamin. They were able to stay afloat for 62 minutes. Clearly, the extra pantothenic acid helped them cope with the stress of swimming in cold water.

But the experiment gets even more remarkable. The last two groups of rats were surgically deprived of their adrenal glands. Group four was given the "RDA" for pantothenic acid. Group five was given "excess" amounts. Group four was able to swim for 19 minutes. But group five stayed afloat for 37 minutes. *Pantothenic acid so boosted the body's ability to handle stress that the animals who had no adrenal glands, but who received bonus amounts of the vitamin, were able to swim longer than the normal animals who received only standard amounts.*

The same researchers ran similar tests on people—without removing their adrenal glands, of course—and found the results to be similar.[3]

B₆ (pyridoxine): Without an adequate supply of pyridoxine, energy cannot be produced in the cells. Pyridoxine is required for the metabolism of carbohydrates, fats, and proteins. B_6 will not cause a toxic reaction if taken along with the rest of the B complex.

I have found in my own practice that pantothenic acid and pyridoxine help the body suppress the effects of short-term flare-ups of inhalant allergies, like hay fever. The doses I use are as follows:

Pantothenic acid, 500 mg. three times a day

B_6, 500 mg. three times a day

NOTE: These doses are most effective when used on a short-term basis, therapeutically, for short-term allergic reactions such as hay fever—not for daily, long-term use.

These dosages, taken for one month or less with the other maintenance supplements, and the rest of the B complex in particular, will not typically cause trouble. You'll know within three or four days whether these supplements are going to work. If they don't work in that space of time, discontinue them and reduce the dosage to your normal maintenance levels.

Folic acid: This B vitamin is probably the most under-supplied nutrient in the United States. Most people don't get enough. "Friendly" intestinal bacteria can manufacture folic acid. But so many people take antibiotics without bothering to replace the destroyed intestinal flora, this source of the vitamin is frequently cut off from them. One of the primary sites a folic acid deficiency shows up is in the red blood cells, where it creates a specific type of anemia. Often, the first symptom of folic acid deficiency is weakness. Drowsiness, irritability, diarrhea, and depression follow.

B_{12} (cobalamin): A deficiency of B_{12} inhibits the production of red blood cells and creates a specific type of anemia, pernicious anemia. As in most B vitamin deficiencies, the nervous system is also affected. Symptoms of B_{12} deficiencies include weakness and drowsiness.

Intrinsic factor, a digestive substance made in the stomach, must combine with B_{12} before the vitamin can be absorbed. Most people don't have a problem absorbing B_{12}. However, if you are regularly experiencing digestive upset, such as gas and bloating, there's a chance that you're having a problem manufacturing enough digestive enzymes —and probably also intrinsic factor. If this is the case, you may try the sublingual form of B_{12}, which is absorbed directly into the bloodstream. If that doesn't work, you should consult your doctor about the possibility of B_{12} injections.

Others: Biotin, choline, inositol, and PABA (para-aminobenzoic acid) are other B family nutrients that should be included in a balanced B complex supplement. Biotin is required for carbohydrate metabolism and the maintenance of the thyroid and adrenal glands. Choline is required for the proper function of the brain, nervous system, and liver. Inositol is important to nerve conduction. PABA is an antioxidant: it protects the red blood cells from ozone.

Vitamin C

Vitamin C may well be the single most important vitamin when it comes to rebuilding our adrenal glands and our energy supply. It's no accident that the adrenal glands are the site of the body's highest concentrations of vitamin C. The vitamin is crucial to the maintenance of healthy structure and function of the glands. Whenever we are challenged by stress, vitamin C is mobilized by the adrenal glands and from other temporary storage sites in the body. The symptoms of vitamin C deficiency mirror those of adrenal insufficiency: fatigue, weakness in the muscles, decreased tolerance for stress, blood sugar imbalances, and tendencies to express allergies.

Vitamin C is a powerful antioxidant, and it plays an important role in cellular respiration. All forms of stress or exertion, including exercise, raise our requirements for vitamin C. When muscles are exercised, they use up vitamin C at a faster rate. This may help explain why one of the first signs of vitamin C deficiency is weakness.

How much vitamin C do we need? For most people, the appropriate dosage is only 2–6 grams a day, in powdered form, dissolved in diluted fruit juice or water. The therapeutic dose I recommend for someone recovering from fatigue who needs to rebuild the adrenal glands is 8–12 grams a day of a buffered vitamin C powder. Vitamin C powder made from sago palm is recommended for very sensitive individuals, because it will minimize any tendency to sensitivities, which may show up with other forms of vitamin C powder. However, if you know you're not sensitive, you can use one of the other ascorbate powders discussed in Chapter 5. Take a level teaspoon three times a day in a large (12-ounce) glass of water or diluted fruit juice (half

water, half juice). Taken as a drink in this way, vitamin C
should be consumed between meals, at least one hour before or two hours after meals. Don't drink liquids with a
meal. They dilute the natural digestive enzymes and can
cause uncomfortable symptoms of indigestion.

Is 8–12 grams a day of vitamin C toxic? No. If you take
more than you need, no harm will result. Loose bowel movements and gas may occur—and that can serve as your sign
that you've had too much. If that occurs, you should not take
any more vitamin C for at least 12 hours. Then you can
continue with half a teaspoon two or three times a day. Keep
adjusting the dose downward so that you can take as much
as possible without experiencing a lot of gas or loose bowel
movements. First thing in the morning is an excellent time
for a vitamin C drink. It will wake you up immediately and
also help to promote a bowel movement.

A few years ago, it was claimed that vitamin C inter-
feres with B_{12} metabolism. This claim was discovered to be
false, the result of a laboratory error.

You might also hear the claim that excess amounts of
vitamin C can cause kidney stones. This is an extremely rare
occurrence. If you have a history of gout, a high uric acid
level, and/or kidney stones, you can greatly reduce any re-
mote possibility of vitamin C's producing stones (or of developing stones in general) by simply taking extra magnesium.
Magnesium oxide in daily doses of 200 mg. will help you
avoid kidney stones.

MINDING YOUR MINERALS

Minerals are very important to the rebuilding and maintenance of your energy supply. I generally recommend that you
find a good multimineral supplement, one containing all the
minerals in appropriate amounts and proportions. If you
need more of certain minerals, such as calcium and zinc,
individual forms of these are also readily available. Once
again, I recommend that you divide the following doses between two meals.

The minerals that should be in your mineral supplement include:

Zinc: Although zinc may be the last mineral alphabetically, it comes first when you need to heal and defend your body against illness. Zinc is a vital rebuilder and booster of the immune system. It is especially helpful in healing the pancreas, which has most likely been ravaged by a high-sugar diet. And because zinc is required for the production of testosterone, lack of libido and impotency may be a result of zinc deficiency. I recommend a daily supplement of 15–30 mg. zinc.

Calcium: Most of the calcium in the body is used to maintain the bones and teeth. However, a small amount of calcium is used to maintain fairly constant blood levels of the mineral, which are used to control muscle and nerve excitability. A deficiency of calcium may result in loss of the mineral from the bones, causing progressive weakening of the bones, a condition called osteoporosis. But it can also result in heightened irritability in the muscles and nerves. Convulsions can occur if the drop in blood levels of calcium is severe enough. But before that happens, the muscles can merely become easily fatigued from a normal workload. Your supplement should contain 600 mg. of calcium per day. Women over 50 or past menopause should take at least 1,000 mg. of calcium every day. If your multimineral doesn't supply near this amount, you can get more calcium through an individual supplement.

Chromium: The so-called "glucose tolerance factor" form of this mineral is very important to balance out the blood sugar. Chromium is a metabolic partner of insulin. Without insulin, the appropriate cells cannot store glucose for later use. Without chromium, insulin is less effective. Consequently, fatigue and other symptoms of low blood sugar (irritability, depression, etc.) occur sooner. Your multimineral supplement should contain about 400 mcg. (micrograms) of chromium.

Copper: This mineral is necessary to balance out the zinc and also to maintain the nervous system. Zinc-copper ratio is important to the regulation of thyroid function, which is all important when it comes to regulating metabolism and energy supply. Your supplement should contain 1 mg. of copper.

Iodine: Iodine helps keep the metabolism on an even keel by maintaining appropriate function of the thyroid gland. A good multimineral supplement will contain from 100 to 200 mcg. The best natural source of iodine and all other trace minerals is kelp and other sea vegetables (seaweed).

Iron: Iron is necessary for the production of hemoglobin, which actually carries the oxygen in the red blood cells. One of the first—and by far the most common—symptoms of iron-deficiency anemia is fatigue. About 25 mg. of iron a day is plenty, unless you're pregnant, nursing, heavily menstruating, have recently undergone surgery or had a bleeding injury—in which case you'll need about twice as much. A good tip on taking iron supplements: Vitamin C enhances the absorption of iron, so it's a good idea to take vitamin C with your iron supplement.

Magnesium: Magnesium is a metabolic partner in several vital metabolic functions. Of particular interest to those rebuilding their energy supply is the fact that magnesium is necessary to the metabolism of calcium and B vitamins.

Magnesium has been used in conjunction with potassium as a supplement to combat what one physician has referred to as "housewife syndrome." The symptoms, as he describes them, include: "fatigue in the morning after what should have been an adequate night's sleep, and a lassitude that, although less severe in the morning, rapidly increases during the day, with utter exhaustion by nightfall." These symptoms, along with vague pains, headaches, insomnia despite sleepiness, and lower back pain, were found in men as well as women, despite the name of the complaint.

Regardless of the name, a combination magnesium-potassium aspartate supplement successfully banished "housewife syndrome" in 87 percent of the patients in one study. I have found a combined magnesium-potassium aspartate supplement to greatly aid many of my patients in regaining lost energy.

I recommend that your mineral supplement contain at least 400 mg. of magnesium.

Manganese: This mineral also plays a role in insulin's balancing of blood sugar levels. Manganese is also vital to the integrity of the tendons and ligaments, and so is a very good supplement to use when recovering from an injury to these parts of the body. Your daily mineral supplement should contain 15 mg. If you are recovering from an injury, you should increase your daily manganese to 25 mg.

Potassium: Neuromuscular irritability and fatigue, mental disorientation, and cardiovascular problems are all results of potassium deficiencies. Potassium was part of the combined mineral supplement given to people suffering from "housewife syndrome," described above. As mentioned above, the aspartic acid mineral salt of potassium and magnesium has helped many of my patients overcome fatigue. With several solid references to the energy benefits of potassium-magnesium aspartate supplements appearing in the literature, I'm surprised more physicians and lay people don't know about this wonderful substance. Your daily mineral supplement should contain 200 mg. of potassium. You may also take a potassium-magnesium aspartate supplement in recommended dosage in addition to your multimineral.

Selenium: Selenium is a very important antioxidant, especially in combination with vitamin E. It's also recognized as extrememly important to the health of the immune system and resistance to cancer. Your daily mineral supplement should include 300 mcg. of selenium.

Sodium: To many people, sodium is a dirty word. But though most people should not use sodium salt because of

RECOMMENDED DOSAGES FOR VITAMINS AND MINERALS
[TO BE TAKEN IN TWO DIVIDED DOSES EACH DAY]

	THERAPEUTIC	MAINTENANCE
A + beta carotene	20,000 IU	20,000 IU
C	8–12 grams	2–6 grams
E	600–800 IU	400 IU
B_1	200 mg.	100 mg.
B_2	100 mg.	75 mg.
B_3	100 mg.	50 mg.
B_5	800 mg.	400 mg.
B_6	400 mg.	200 mg.
Folic acid	300 mcg.	100 mcg.
B_{12}	1,000 mcg.	500–1,000 mcg.
Choline	100–200 mg.	75 mg.
Inositol	200 mg.	75 mg.
PABA	200 mg.	75. mg.
Biotin	200 mcg.	75 mcg.
Zinc	30 mg.	15 mg.
Calcium	600 mg.	600 mg.
Chromium	400 mcg.	200 mcg.
Copper	1 mg.	1 mg.
Iodine	100–200 mcg.	100–200 mcg.
Iron	25 mg.	25 mg.
Magnesium	400 mg.	400 mg.
Manganese	15 mg.	15 mg.
Potassium	200 mg.	200 mg.
Selenium	300 mcg.	300 mcg.

its effect on blood pressure, it's good to keep in mind that sodium is the major mineral in the fluid surrounding the cells and that it is lost in sweat. Sodium deficiency is rare. But if you sweat a lot and eat a low-sodium diet, it is possible to be deficient. Symptoms of sodium deficiency include headaches, muscular cramps, and weakness. Unless you have high blood pressure, the discreet use of small amounts of sea salt is not inappropriate. Sea salt is always to be preferred over table salt because in the former, sodium chloride is accompanied by a naturally balanced complement of trace minerals as they exist in seawater, which sustains life. Potassium salt is fine, because most people could use a little more potassium anyway. And there is evidence that a beneficial ratio of potassium to sodium helps to control blood pressure.

Trace minerals: Trace minerals such as vanadium, molybdenum, and others are sometimes found in high-quality multimineral supplements. Usually, such a supplement is made from volcanic residue or from dry, ancient seabeds with highly mineralized sea vegetables. You may want to find a good trace mineral supplement that contains these extra substances. Precise dosages are not important. If your supplement contains enough of them so that they are listed on the label, that will be sufficient.

OTHER SUPPLEMENTS THAT BOOST YOUR ENERGY

Many people feel that the only important supplements are vitamins and minerals. Actually, several other supplements are vital as steps to health and high energy.

Enzyme Supplements for Digestion

If you have a digestive problem—belching, burping, gas, bloating, and generally an uncomfortable feeling after eating—this is often a signal that you are not digesting your foods properly. (It can also be a sign of a food sensitivity, so first rule that out.) The symptoms alone are annoying enough, but the more important reason to take care of this problem is that if you are not digesting properly, you're not getting the full benefit from either your food or your supplements. The supplement recommendations I am going to make will help these conditions, whether incomplete digestion is the result of a lack of enzymes or a sensitivity to the particular food itself.

Gas usually means that food is passing through the bowel before it's completely digested, so bacteria can thrive on the greater than normal amount of food and produce gas. Generally, protein foods or the protein parts of foods are the greatest offenders. But complex carbohydrates cause the problem, too, mainly because people don't chew them well enough or they drink with their meals. Drinking liquids with a meal dilutes the digestive enzymes and lessens their ability to digest food. That glass of ice water you drink with your meal may be the lone culprit causing your indigestion and poor assimilation. Chewing your food completely and not drinking with your meals may solve all or part of your problems with incomplete digestion. However, you may still need to take some supplements.

Of the digestive enzyme supplements, the most important is extra hydrochloric acid (HCl), commonly available as betaine hydrochloride. (This should never be taken by anyone who has had a history of ulcers or gastritis. Check with your physician if you have any doubt.) Most people with indigestion need *more* acid, not less. Additional acid in the stomach can improve digestion remarkably. I've always found it interesting that the medical profession is so eager to pass out antacids, which often get rid of the gas and bloating symptoms but compound the basic nature and cause of the problem.

Pancreatic enzymes are also a good idea. They bolster the supply of the body's own pancreatic enzymes, which aid in the alkaline digestion taking place in the small intestine. Complete digestion of proteins, fats, and carbohydrates depends on pancreatic enzymes. In addition, by assisting in protein digestion, both HCl and pancreatic enzymes minimize food sensitivity reactions.

If you have a particular problem digesting fatty foods —which I recommend you limit in your diet anyway—or if you've had your gall bladder removed, then you need extra help in the fat-digesting department. The cardinal symptoms of this are belching and burping. You should be taking a bile factor supplement, which contains ox bile and other accessory digestants.

HCl supplements usually come in standard doses. Follow the directions on the label and take them toward the end of a meal, never on an empty stomach. They will shut down the body's own natural attempt to make hydrochloric acid if you take them at the beginning of a meal. Also take enterically protected pancreatic enzymes and/or bile factors near the end. "Enterically protected" means that the stomach acid will not neutralize them before they get to do their work. Take them in the standard doses recommended.

These digestive enzymes should be taken until you've gotten rid of your food sensitivities, until you've rebuilt your diet along my recommendations, and until your digestive abilities are restored and your symptoms start to go away. As your symptoms go away, you can experiment with reducing or eliminating the enzymes. Many people no longer need them after a few months because their systems have totally rebalanced.

Essential Fatty Acids

Essential fatty acids, linoleic and arachidonic (which can be synthesized from linoleic) play a role as hormone precursors and in cell membrane structure and integrity. Most quality diets have enough of these. But low-fat diets, like the Pritikin diet in particular, may be deficient in them. Since low-fat diets are desirable to prevent degenerative diseases, and in this book I have oriented you toward a low-fat diet, I recommend that a small amount of essential fatty acids be taken as a supplement to ensure that they be there for what they're needed for. The simplest way to do this is to incorporate *small* amounts of olive, sesame, safflower, or sunflower oil into your diet.

Amino Acids

I recommend free-form amino acids. "Free-form" means they have been chemically separated from each other and are directly absorbable. Though they are supplied together in one tablet or capsule, they are biochemically distinct. I'm not talking about protein powders or predigested protein. Free-form amino acids are usually available in capsule form, generally in doses ranging from 500 to 800 mg. Make sure you get a balanced blend, not just one amino acid or another.

For many people suffering from fatigue, a free-form, balanced amino acid supplement is very helpful during the rebuilding phase. The hardest type of food for most people to digest is protein, and when protein digestion is poor, the essential amino acids may be poorly assimilated. Since they're so important to all aspects of metabolism, it's vital to get them all in order to have high energy.

In order to get meaningful results, you need to take 10 to 20 amino acid capsules two or three times a day between meals. You don't want to take these with meals because then they compete with the protein in the food and tend to work against one another. Experiment with the number of capsules and the time at which you take them. Late morning and late afternoon frequently are the best times.

Take these supplements with diluted fruit juice or fruit. The natural sugar will produce an insulin release that helps the amino acids to be absorbed into the appropriate cells more readily. Usually within one half to one hour you'll get an obvious lift and notice that you have more energy. You will feel better right away, and that's part of the benefit of these supplements.

You probably won't need these supplements (which are usually expensive) after the rebuilding phase.

I generally do not recommend that people take single amino acids. But the following are two exceptions to that rule.

Tryptophan: Some people have difficulty getting to sleep and staying asleep. This, of course, can adversely affect their energy levels. If you have trouble with insomnia, please try to stay away from sleeping pills. Sleeping pills may put you to sleep, but they do not provide the deep, natural, restful sleep the body truly needs to re-energize. Tranquilizers and sleeping pills can produce fatigue for many days after they're taken. If you're using these drugs, it's a major step to abandon them, but it's a step you'll eventually have to make if you want to rebuild your energy supply. However, do not discontinue any prescribed drug without first consulting your physician. Following all the other cleansing and rebuilding suggestions in this book will also help decrease the need for these types of drugs.

Instead of drugs, one of the things you can try is reading before you go to bed. Or try some form of meditative breathing or meditation to slow down and "turn off" the mind. Tryptophan, a single free-form amino acid, in doses of 500 to 1,000 mg., can be used with 50–100 mg. of B_6 (which helps its conversion to serotonin, the brain hormone that promotes sleep) to help you get to sleep and stay asleep. It should be taken with fruit juice half an hour before bed. The juice stimulates the release of insulin, which helps the tryptophan achieve its desired effect more readily. Most protein foods contain tryptophan. However, when tryptophan is taken with other amino acids, which are present in the food, its effects are minimal. A glass of warm milk, the traditional bedtime drink, does contain tryptophan. But whether there is enough to have the same effect as a dose of the single amino acid is questionable.

One of the benefits of tryptophan is that it does not create a hangover or a drugged-out feeling the next day. Tryptophan can also be taken during the day in the therapeutic period if you're a nervous, edgy person. If you've been taking a lot of minor tranquilizers during the day to stay calm, you can try substituting tryptophan and thereby begin to cleanse the tranquilizing drugs out of your body.

L-glutamine: Glutamine can reduce the craving for alcohol and sugar. I've seen a lot of people in my practice reduce their craving for sugar by taking glutamine, 500 mg. twice a day between meals, with diluted fruit juice. If you have an uncontrollable urge for sugar, it will help you reduce it. Generally, you will find that the craving is due to some food sensitivity. If you follow the instructions in the detox and desensitization steps and find out what the sensitivities are, you will usually knock out the craving for sugar. But if your first detox does not completely remove the craving, glutamine will help.

Special Healing Foods

Certain foods are particularly helpful because of their strong potential for healing and building strength. These foods are usually high in vitamin-mineral content. They also have healing properties that seem to go beyond the impact usually associated with their particular nutritional components. If you are not allergic or sensitive to them, they can be very beneficial. They are a bit unusual, so bear with me as I describe their benefits.

Calf's liver: Calf's liver is high in protein, B vitamins, vitamin A, and minerals. When taken raw, it also appears to have a special healing effect on the liver and the pancreas.

It's important that the liver be *organically raised.* The liver is responsible for detoxifying most of the poisons that come into the body, and if the animal is not organically raised, the liver will contain all the poisons and hormones that have been fed to the animal in the commercial feedlot. In a clean, organically raised animal, the liver provides extra vitamins, amino acids, and other nutritional components, without the hormones and other chemicals. Calf's liver is preferable to the liver of a full-grown animal because it has not had as much time to accumulate toxins.

I recommend raw calf's liver to people who are anemic to build up their iron stores. One of the ways it can be made palatable is to take about 2 ounces of raw calf's liver, chop it up, put it in a blender with some tomato juice, add a little vegetable seasoning or other seasoning that you like, blend it, and drink it down. This tastes much better than you think it will. You'll be able to get it down just fine, unless you've made up your mind that it's going to taste terrible.

The raw liver should only be necessary during the rebuilding phase. However, if you want to continue it, that's up to you. If you have a cholesterol problem, however, liver should be used sparingly.

Brewers' yeast: This food is used by many people as a source of natural B vitamins, protein, selenium, and chromium. Many people find that a teaspoon of brewers' yeast twice a day between meals, taken in a drink or mixed with food, helps increase their energy and makes them feel better. However, the yeast is a highly allergenic food. In sensitive individuals, it can actually maintain the craving for sensitivity-offending foods and produce both fatigue and a clouded, fuzzy-brain feeling.

Acidophilus: Antibiotics, which are taken wholesale in this country, wipe out the friendly bacteria in the bowel. These friendly bacteria not only prevent constipation and keep us from getting infections, but they also manufacture some of the nutrients that we need, like folic acid. After a detox program, after taking antibiotics, or if you have continual problems with your digestion or constipation, you should take acidophilus. I recommend one of the new cultures with extra growth potential, developed in the last few years. Since most acidophilus is raised on a milk-based culture, if you are allergic to milk, you should try to find a non-milk-based acidophilus. Take the dose recommended on the label.

A final word about supplements: Do not expect them to perform the entire job of rebuilding your energy supply. Supplements are only important links in a much larger chain of good nutrition and self-care. All the other links must be in place before the chain can be as strong as you want it to be.

Steps to High-Energy Supplementation

Step One: Some people can jump right into a full-fledged supplementation program, and some people need to ease their way in a little at a time. If you are among the latter group, your first step can be to begin to use a high-quality multivitamin-mineral supplement. The doses should be close to the doses I mentioned in this chapter.

Step Two: You may not notice an appreciable difference at Step One. In that case, for your second step you should add higher amounts of vitamin C and E, plus some digestive enzymes, to your supplement program. You may also wish to add a potassium-magnesium aspartate supplement.

Step Three: For this step you should endeavor to incorporate all the supplements from this chapter into your program. You may also wish to add amino acids and any of the special healing foods that are appropriate.

HIGH-ENERGY TIP

TAKE VACATIONS

When you get a sense of starting to get a little burned out or run down, or when you don't enjoy your work anymore, it's time to take time off. How often or for how long you need to do this is unique for every individual.

Take breaks. It doesn't have to be a week or 10 days or a month vacation. You can have a weekend vacation, getting out of town away from your usual surroundings to relax and break up the monotony. Go by yourself if you need to be by yourself. There's nothing wrong with needing a bit of free time.

You can have a evening's vacation as well, during which you mentally sort things out or do something interesting, rewarding, or entertaining for your sense of self.

When you take that three-day vacation or that week off, or even that evening, stop and get a sense of how you feel. The comfort, the relaxation, the clearheadedness, the lack of burden upon the mind is a wonderful feeling. It's the reason people love to take vacations. And remember, a vacation is a state of mind you can take home with you. When you come back to your daily routine and work, carry the vacation mentality back with you. You will find that you have more peace, more time, and more energy going through your daily routines.

8

Exercise for High Energy

"Each time you exercise, you come back stronger. Before long, you flat out get tough, both mentally and physically."

OLYMPIC DECATHLON
CHAMPION RAFER JOHNSON

Using energy creates energy. Mike, one of my patients, is living proof of this. Mike is 73 years young. He thinks of himself as young and always manages to think positively. He has a pretty good diet and has come to me from time to time for some fine-tuning with his supplements and other health programs. Mike either rides his bike or runs five miles a day. He'll tell you that as long as he maintains his regular exercise he functions very nicely. Exercise doesn't take energy away from Mike, it creates more energy for him.

When you're fatigued, one of the last things you're inclined to do is go out and exercise. You say to yourself, "Oh, I'm tired, I don't want to go out there, I just want to stay home and take a nap." What you have to realize is that sometimes you have to force yourself to go out and exercise through the fatigue.

This is especially true for the person with food sensitivities, who isn't actually physically tired but is perceptually tired because the brain is fogged from the food sensitivity reaction. You should always be aware of this and encourage yourself, whenever possible, to break through that illusion of fatigue. Many times if you focus on what it is you want to do, the energy can be created to do it.

Many of my patients have had the experience of feeling tired and then forcing themselves to exercise—and feeling wonderfully alive and refreshed after the exercise. If you were truly physiologically fatigued, there is no way you could force yourself to go out and exercise and come back feeling better. When the fatigue is just a perceptual problem, however, you can clear that out by stimulating the metabolism and oxygenating the organs and the brain. Often you can literally run (or cycle, swim, or walk) away from your fatigue.

People always have good reasons why they shouldn't exercise: they're too busy, don't have enough time, they're tired . . . etc. A busy person is never going to find the time. *You have to make the time!* You've got to make that commitment to yourself. Know that when you come home from work you're going to stop at the gym. Or decide that you're going to get up half an hour earlier to exercise.

Consider exercise as basically important to your life as

eating and sleeping. You wouldn't consider *not* doing those things on a regular basis.

A program of regular exercise three to five times a week is as essential to freedom from fatigue as eating and sleeping. Exercise may sound like work, but it's what the body thrives on. Without it, our health and vitality suffer.

OF COBWEBS AND ENDORPHINS: THE BENEFITS OF EXERCISE

Work—defined as a unit of energy—actually makes the body better. Here's exactly what happens:

Your whole circulatory system becomes more efficient. Exercise conditions the heart so that it pumps the blood with fewer beats per minute, therefore expending less effort to do the necessary amount of work.

The amounts of the damaging type of cholesterol (LDL) and fats in the bloodstream decrease, and the protective type of cholesterol (HDL) increases.

With regular exercise, the small blood vessels (microcapillaries) stay viable longer in life, resulting in better oxygenation of the tissues. Since oxygen is the most important nutrient, this postpones the inevitable degenerative changes of aging (especially if you're taking the recommended supplements) and maintains the healthy structure and function of the organs and glands, especially the heart, lungs, and liver.

For all of these reasons, far fewer heart attacks and other degenerative diseases occur among people who exercise regularly.

Of particular interest to the person experiencing chronic fatigue, exercise tends to normalize the balance of adrenal hormones. This conditions the adrenals and fortifies them so that more severe stresses can be adequately handled.

Physically fit people *look* better, too. They have more muscle and less fat. Their posture is better. Their skin retains its youthful elasticity longer. And they rarely have a weight problem, because exercise not only burns up calories, it can also help raise the basal metabolic rate and control the appetite.

Exercise combats chronic fatigue and listlessness by normalizing blood sugar levels and encouraging the burning of body fats for energy. When the body is physically conditioned (and not abnormally reacting from a direct cellular sensitivity to food), energy is normally available when there is a need for it—without resorting to some kind of abnormal and stressful stimulant like caffeine, nicotine, or a sensitivity-offending substance to pursuade the adrenals to liberate some energy.

We can also use exercise to clean the cobwebs out of the mind for clearer thinking, to help alleviate emotional and spiritual pains, and to promote a feeling of inner strength, capability, and well-being. Exercise channels tension, frustration, and other pent-up energies out of the body and mind, leaving tranquility.

We are only beginning to discover how exercise does all these things. One recent discovery is the presence in the body of hormones that actually promote feelings of peace and happiness. When you're feeling contented and happy, certain chemicals in your brain are helping you feel that way. The most important of these is a class of chemicals called endorphins. Endorphins are the body's natural pain-killers, but they go even further than alleviating pain. Endorphins can create an intense inner feeling of happiness and well-being, a totally natural and healthy "high." Blood levels of endorphins are increased by exercise.

Exercise seems to transcend the realm of the physical body. Runners often call it "exercise high" or "third wind," because it sets in after they've been running for several miles. Those who experience it feel close to the spirit, likening it to mystic states like meditation or Zen, or to an out-of-body experience.

Although there is speculation that the endorphins and other hormones may have something to do with this transcendent experience during exercise, all we know is that exercise does seem to put us in touch with a very powerful, beneficial aspect of our own existence.

Through regular exercise, you can begin to create a ladder out of the pit of fatigue, and you can keep on climbing higher, to new levels of health and vitality. In particular, as the adrenal glands are strengthened by regular exercise, you can look forward to a gradual freeing from the need to continually jolt the adrenal glands for energy.

As Rafer Johnson says in the quote at the beginning of this chapter, you will "get tough." You'll become so strong, so tuned in to the way your regular exercise makes you feel, that you'll allow few things to stand in the way of your getting out there and doing it.

FIRST, CHOOSE YOUR SPORT

It's vital that your exercise be fun. So even before I give you tips on getting started, I suggest you choose an aerobic activity that you enjoy. Otherwise, you may not have the motivation necessary to stick with it.

Your first consideration might be whether you want to exercise indoors or out. Many people prefer to exercise indoors because it solves the problems of weather, bugs, dogs, security in the neighborhood, and a tight schedule.

There are plenty of ways to exercise inside your home. Yoga, for example, is tailor-made for warm-ups. Then you might go on to easy running in place. Wear running shoes and jog on a piece of very soft carpet or on a mini-trampoline in order to avoid joint and ligament stress. Let your arms swing at your sides and lift your feet at least 6 inches off the floor.

You can also take a home class in aerobic exercise by following along on one of the TV exercise classes or videotapes.

A stationary bicycle provides an ideal form of indoor exercise. You can set the tension to give a more or less strenuous workout. It's easy on your joints, provided the pedals are adjusted so that when the pedal is at its lowest position, your leg is *slightly* bent at the knee. As you rack up the miles on the odometer, your sense of satisfaction will motivate you. And if you find it boring, you can always place the bike near a TV set.

Jumping rope is a vigorous activity suitable for someone in good physical condition. You can do it just about anywhere that has a high ceiling and a sturdy floor. But you should not do it until you are already in good shape.

Rowing machines also are excellent. They provide a good aerobic and resistive workout at the same time.

Even if you prefer to exercise outdoors, it's a good idea to have some type of preferred indoor exercise to serve as a backup during inclement weather.

If you exercise outdoors, choose your spot carefully. Take your daily activity in as pretty a setting as you can find —perhaps near a lake, on a beach, or in a park. An aesthetically pleasing setting will help create a sense of peace, which reduces stress and helps rebuild the adrenal glands.

Clean air is also important. You'll be breathing at several times your sedentary rate, so you don't want to be jogging down a busy street or pedaling alongside a freeway at rush hour. The air is usually cleanest early in the morning. Parks, beaches, and residential neighborhoods have cleaner air than busy downtown areas. Check with your local Air Quality Board for information about the cleanest places and times.

Walking is a great outdoor exercise, especially for those who are not physically fit and for older people. A brisk walk may be nearly as good exercise as a slow jog, and a lot easier on your joints. All you need is a comfortable pair of shoes. Find a place to walk that you like, a path that's interesting and pleasing. If you find yourself getting bored, set out in new directions. Or find a friend to walk with you.

If you like golf, it's possible to get five miles of walking under your belt between the first hole and the 18th. But if you wind up riding the golf cart instead, you might as well be watching TV for all the exercise you'll be getting.

As you work your way up to new levels of fitness, you may want to try more vigorous activities such as skating (roller or ice), swimming (*real* swimming—laps or distance, not splashing around), bicycling, rowing, or skiing.

Besides aerobic exercise, there are two other kinds of

exercise: stretching-flexibility exercises, like yoga; and resistive exercise, like weight training. Of the three, aerobic is the most important and vital. Flexibility exercise comes next in importance. Alone or combined, flexibility and weight training are not adequate replacements for aerobic conditioning.

As we get older, there is a natural tendency for muscles to tighten, tendons to shorten, and for our bodies to generally lose elasticity. Stretching exercises loosen up the body and facilitate good circulation and prevent injuries. Feeling your muscles slung loosely on your bones rather than pulled taut around numerous tension centers helps give a youthful, happy feeling that promotes calm and tranquility and the rebuilding of health and vitality.

I'm a big fan of yoga. I recommend taking a course in it and doing it regularly. But if your time is limited, you might just incorporate some extra stretching exercises into your cool-down period (the 10-minute period immediately after your aerobic exercise).

As for resistive training, with the increased popularity of "pumping iron" at health clubs, I believe it is important to get a few facts straight. Resistive training involves relatively short, rapid, forceful movements such as isometrics, lifting weights, and working out with mechanical apparatus in order to build strength and size in the muscles. If you want that well-defined muscular look and you have lots of time to put into it—*without neglecting your aerobic conditioning*—go ahead. But let me make two points clear.

First, resistive training should never be done in place of aerobic exercise, only *in addition to it.* If time is a consideration and you have to choose one or the other, choose aerobics. Not only does resistive training do little or nothing to promote cardiorespiratory efficiency, but it actually tends to *impede* circulation during the muscle contraction, rather than enhance it. Arterial blood pressure rises markedly during the contraction and can put an extreme burden on the heart. This effect is the cause of sudden death while shoveling snow. It's also why older persons or those with circulatory or respiratory weaknesses should use extreme caution when dealing with so-called "static" exercise.

Second, resistive training also has a tendency to

shorten the length of some of the muscle fibers and create tightness. For this reason, stretching should be an integral part of body-building, too. You should spend as much time in stretching routines as you do in resistive routines. If you don't have the time to combine aerobics and stretching with your resistive work, then cut out the resistive. Big muscles won't do you much good if you're bound up in knots, and your time would be better spent healing yourself with aerobic and stretching exercises.

DO YOU NEED A STRESS TEST?

As vital as exercise is to vigorous good health, overexertion can trigger heart attacks in individuals whose heart and/or blood vessels are in poor condition. The *fear* of this happening leads many people to postpone getting in shape until they're cleared by a stress electrocardiogram (ECG). Of course, they also put off making the appointment for the stress test, for fear of what they might find out. And while they're sitting in their armchairs, they're sinking deeper and deeper into fatigue and poor physical condition.

You're putting yourself in far more danger by not doing the things your body needs to survive than by doing them. It's not usually the person who takes up gentle jogging, cycling, or walking and gradually conditions his way *out* of an undetected cardiac abnormality who ends up as a coronary statistic—it's the fellow who clings to his sedentary ways until a blizzard forces him to shovel out his driveway.

Still, a stress ECG is a good idea for some people before they embark on an exercise program. Those people include males over 40 and females over 50 who have any of the following symptoms: vague chest pains, irregular pulse, shortness of breath, family history of early heart disease or high cholesterol, sedentary lifestyle, diet high in cholesterol or saturated fats, smoking, obesity, diabetes, high blood pressure, rheumatic fever as a child. The more of these that pertain to you, the more reason you have to let a good cardiologist check you out—especially if you have a tendency to push your body too hard or if you intend to take up a strenuous sport.

At the very least, let your physician know of your exercise plans and get assurance from him that there is nothing in your medical history that spells danger. In the event that there *is,* a stress ECG can help your doctor prescribe a safe exercise program. He may recommend that you enter a medically supervised exercise program designed for coronary patients and those at high risk—at least until your body is sufficiently conditioned.

SOME COMMONSENSE CAUTIONS

The rest of us can play it safe without looking to electronic gadgetry to tell us whether we are fit to run around the block. All we need to do is follow two simple rules.

First, listen to your body. There are signals built into the human mechanism that will tell you when you're overdoing it. *Pay attention to them.* If you are experiencing *pain* (not soreness) in the legs or side, trembling, head pounding, nausea, dizziness, breathlessness, or a very rapid pulse, *immediately slow down (to a walk, if you're running), breathe deeply and stretch the muscles of the rib cage by reaching upwards.* The message here is *not* "Don't do this ever again," but rather: *"Next time, take it slower."*

NOTE: A "suffocating pain" or feeling of tightness in the chest, neck, jaw, back, or left shoulder or arm may be angina pectoris, an indication that your heart muscle is not getting enough oxygen. Since this is a common symptom of heart attack, be sure to see your doctor about it right away.

Second, a workout shouldn't mean all-out exertion. You should keep your pulse well below your heart's maximum capacity (see graph), but still high enough to condition it.

One good indicator of whether you are taking it easy enough is whether you can carry on a conversation while exercising. If you can't, you're probably going too fast. Find someone who is at your level of fitness and exercise together.

If your fatigue has been severe and chronic, it's especially important to take it easy. The person with weak adrenal glands typically feels tired, either because of a blood sugar problem or because of a sensitivity addiction-withdrawal phenomenon. So if you're "addicted" to some sensitivity-offending substance, you may not be inclined to exercise. You may feel you don't have the energy to spare.

In fact, your body, particularly your adrenals, may *not* be able to deal with the stress of heavy exercise. At this point, heavy exercise will actually do more harm than good. It will be just one more stress for your tired adrenal glands to cope with.

So it's even more important for you to start off slowly. Fatigued people should take their exercise dose three or four times a week rather than five or six. This gives the body a chance to recover between sessions. And don't try to cram a week's worth of exercise into one strenuous Saturday morning workout. Besides the danger to your heart, muscles, and joints, any benefits derived will likely be lost over the next six days of inactivity.

At the same time you're exercising, you will be rebuilding your body and adrenal glands using all the other procedures mentioned in this book. Only then will your adrenal glands be ready for more challenging, vigorous physical activities. To initiate a safe exercise routine, start out with walking to minimize adrenal stress.

WARM UP TO YOUR SPORT

You don't want to put on your running shoes or strap your feet to the pedals of a ten-speed, and take off at top speed. That will be hard on your muscles, tendons, joints, ligaments, adrenals, and heart. Before strenuously exercising you should limber up. Stretch your muscles, particularly the ones you'll be using. There is some controversy over the benefits of pre-workout stretching. Some research has discovered that extensive pre-running stretching actually *contributed* to more injuries than it prevented.

I believe that pre-exercise stretching should be brief, casual, and easy at most. Rather than going through an elaborate series of stretches to warm up, start off your exercise slowly for the first 5 or 10 minutes. Let your body warm up to "operating temperature." If you want to go through a series of stretches, do it after those 5 to 10 minutes. You're better off stretching when you're already warmed up.

NOW FOR THE HUFFING AND PUFFING

Lest anyone take advantage of my warning not to overdo, let me define exactly what I mean by exercise. I don't mean sitting on your living room floor watching an exercise show on TV, doing 10 leg raises, 10 toe touches, some arm circles, and then walking to the kitchen for a snack during the commercial.

What I am talking about is *aerobic* exercise, or rhythmic physical activity utilizing large muscle groups and maintained over a period of at least 20 minutes at a fast enough pace to substantially elevate the breathing and pulse rate. Unless it's elevated beyond a certain level, there will be no conditioning effect.

One indication that your pulse is sufficiently elevated is huffing and puffing. You can't do aerobic exercise and still breathe quietly through your nostrils. For a more accurate indication, you need to calculate your *maximum* heart rate (unless it's already been ascertained by a stress test). You can do this simply by subtracting your age from 220. If you're 50 years old, your maximal heart rate would be about 170. Get a watch that shows seconds. Using your index and middle fingers, take your pulse at the point where your thumb joins your wrist. Count the beats for 10 seconds, then multiply the number by 6 to get your heart rate per minute. Then check Table 8-1 to ascertain your target zone.

If you haven't been exercising lately, or if you have doubts about your cardiovascular or respiratory health, start your exercise program using *half* your maximal heart rate as your target pulse. That would be 85 beats per minute for a

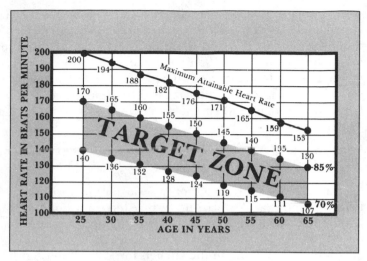

Target Heart Rate Zones

Table 8-1

50-year-old. If you have no problems at this level of exertion, gradually increase to 60 percent. This will be enough to bring about gradual improvement in your cardiovascular-respiratory function.

By maintaining this heart rate for 20 to 40 minutes regularly, you'll soon find you can exercise a little faster or harder without your pulse increasing. This is one sign of the "training effect" of aerobic exercise. At this point you can increase to 70 percent and then 80 percent, on a continuing, gradual basis. Once you are conditioned, you will regularly exercise in the 70–85 percent range.

Resist the temptation to get competitive and push yourself too hard. By taking it nice and easy, you'll avoid a lot of aches and pains that would slow down your progress in the long run and might even cause you to give up exercising.

COOLING DOWN COMFORTABLY

The end of your exercise session should look like the beginning, only in reverse. Cool down by slowing down for the last 5 or 10 minutes. Then you may go through a more definitive stretching routine. Cooling down and stretching are very important at the end of your exercise routine. You want your heart and body to adjust to normal workload and your muscles, joints, ligaments, and tendons to stretch in order to prevent stiffness and pain, and to enjoy the feeling of a loose, strong body. A cool-down/stretch routine is more effective than warm-up stretching and is less likely to produce injury, since the body's tissues have been warmed and loosened by the exercise.

Exercise gives you more energy than it takes away. If you do everything else recommended in this book but don't exercise, you won't get back 100 percent of the high energy that is your birthright. Exercise carries you out of the realm of feeling good and makes you feel great. It makes you feel tough—so you can do anything you really want to do. That's high energy.

Steps to Exercising for High Energy

Step One: Unless you have been bedridden with a serious illness or have symptoms of heart disease, you don't need to get your doctor's permission to take a walk. If exercise is a new concept for you, walking around the block is a way to ease into it. The great thing about walking is that it can be integrated into just about any other activity. You can take along binoculars or a camera on a walking expedition. You can walk from store to store and window-shop. You can park your car or get off the bus several blocks from your destination. If you start off walking, you should measure your progress toward fitness in terms of your pulse, the distance walked, and speed. As you become more and more fit, your pulse will not increase quite so much with equal amounts of exertion. That's a clue that you are getting stronger and can now increase your workout. You might start with walking around the block, but to make progress, you will need to either increase your distance or walk faster. Or you can simply walk more often. Instead of walking around the block three times a week, do it five or six times a week.

Step Two: Increase the speed and/or distance of your walk, by walking briskly one-half to one hour daily. Or choose another aerobic exercise and do it for 20 minutes three or four times weekly. Incorporate some stretching routines into your exercise program.

Step Three: Is exercise a part of your life that you make time for, scheduling other activities so they do not conflict with it? Do you have a sense that exercise is too important to miss or skip? If so, then you are at step three.

HIGH-ENERGY TIP

SLEEP IS HIGH-ENERGY EXERCISE, TOO

If this scene is common in your life, you've got an energy-drain situation that's shocking your adrenal glands and producing fatigue: The alarm jolts you awake at 6:30 A.M. Maybe it's loud or raucous enough to get you out of bed. Or maybe you have to wear out the snooze alarm before you give in and get up. Either way, it's time to change your waking-up ways.

Any unexpected loud noise invokes in any human—even a newborn who has not yet learned specific fears—a fight-or-flight type of reaction. Blood pressure rises—an involuntary response mediated through the adrenal glands, as they respond to any sudden fright. This reaction can occur when you're walking down the street and a car backfires, or when you're asleep in bed and your alarm clock goes off. If the noise occurs on a regular basis, it will contribute to chronic low-level fatigue. Furthermore, the alarm is even more traumatic if it goes off when you're in a period of deep sleep or dreaming.

Sleep is one of the few times in your life when you put your body on "automatic" and trust in its ability to heal, replenish, and rebuild. If you must rely on an alarm clock to wake you, you're not getting enough sleep. You're putting a heavy burden on your adrenal glands to jolt you out of the profound tiredness that results from lack of adequate sleep and other poor care habits. If you continue in this practice, your adrenal glands will not get a chance to rebuild themselves and recover healthy function. Instead, they will continue to tire. Your energy supply will probably not reach the high end of the spiral.

A good night's sleep means that you go to bed early enough so that you wake up naturally when your body signals that it's had enough sleep. You can still use the alarm clock as a backup. But if you go to bed early enough and tell yourself that you want to wake up at, say, 6:30, you can program your "internal clock" to wake you without the commotion of an alarm and disruption of your natural sleep cycle. You'll find yourself waking up 5 to 10 minutes before the alarm goes off. In order to figure out your correct amount of sleep, keep going to bed earlier and earlier until you find yourself waking up naturally at the time you need to get up.

You'll also find that as you get healthier, as your adrenal glands are restored and your energy supply grows, you'll need less sleep.

DO YOU HAVE TROUBLE GETTING STARTED IN THE MORNING?

If you eat or drink late at night and consume foods to which you're sensitive, they'll react on your brain while you're sleeping. Then, if not cleared out of your system by morning, when you wake up you'll find yourself in a brain-fog fatigue state. If you have this problem, you can carry out the detox-desensitization and remove the offending foods from your diet.

You may also feel sluggish in the morning if you have a problem with your adrenals and low blood sugar. Since you haven't eaten anything for the last 8 to 10 hours, you might be in a low blood sugar state in the morning that will also make you tired and dazed. One indication that your blood sugar is low in the morning is if your tiredness is at least partially resolved by eating something. If you have this problem, chapters 4, 5, and 6 are especially important to you.

IS IT OK TO TAKE A NAP?

During the rebuilding phase, it is reasonable to take a nap when you are tired and have the time to do so. Once you rebuild your energy, you really should not need naps. If, however, you have high energy within the framework of your life and you're feeling well, you can help maintain that energy by short naps in the middle of the afternoon. The nap should be just a short, 15- to 30-minute break to help boost your energy, not a desperate attempt to drag yourself out of a state of exhaustion. If you're napping out of exhaustion and fatigue, then you need to investigate other areas to figure out why this is going on.

Some people—Thomas Edison among them—created more energy for themselves by taking naps. Edison was famous for sleeping little at night but taking a lot of short naps during the day to renew his energy.

9

Energize Yourself from Within

"Deep within you is everything that is perfect,
Ready to radiate through you and out into the world.
It will cure all sorrow and pain and fear and loss
Because it will heal the mind that thought these things
Were real, and suffered out of its allegiance to them."

A COURSE IN MIRACLES

By the second year of my practice it was becoming apparent that there was more involved in health than simply detoxing, eating right, taking supplements, and exercising. Many patients implemented these strategies and felt fabulous. But there were too many people for whom this master plan just didn't work well enough.

I began to notice that these two groups of people were different in another way. The first group were those who could take life's crises peacefully. They seemed to have a sense of personal power. They kept on a pretty even keel through marriage or divorce, fortune or misfortune.

The other group let their problems stress them on the emotional-spiritual-psychological plane to the point where their physical bodies were breaking down. They just could not compensate for the stress by simply changing their diet and exercise habits.

A vitally important truth began to emerge about why so many people are tired in general and why so many others are almost completely exhausted: We waste a tremendous amount of energy *worrying* about what's going to happen today, tomorrow, and for the rest of our lives. No single habit is so powerful in its ability to exhaust.

If you have this worrisome habit, you will squander tremendous amounts of physical, mental, emotional, and spiritual energy. Learning to use these types of energies efficiently is as much an art as a science. Furthermore, the ability to live your life at peace with yourself and those around you —the art of being a human being—provides joy, health, and high energy. To teach you to begin to do that, within the confines of our subject, high energy, is the purpose of this chapter.

That may sound like a herculean task. Experience with my patients has shown me that although we certainly cannot provide anyone with all the answers in a couple of sessions, we can "clean the window" so they can look out and get an idea where they need to be going and see the true goals for which they are striving.

This is not too different from improving your diet, your supplement program, and your exercise. Instead of dealing with fruits, vegetables, supplements, bicycles, running shoes, and skis, you'll be dealing with the thoughts and feelings that go through your head and body every day. Thoughts and feelings do not only occupy your mind. They profoundly affect your entire body.

In earlier chapters of this book, I've described how physical or biochemical stresses such as sugar, junk food, and food sensitivities weaken the adrenal glands and drain our capacity for building energy. Nonphysical, mental, emotional, and even spiritual stresses also drain energy by weakening the adrenal glands. Acclaimed researcher Hans Selye, M.D., in elucidating the relationship between stress and physical breakdown, was one of the first medical researchers to link stressed, overworked adrenal glands to states of fatigue and exhaustion.[5] This makes sense, since the fight-or-flight response is evoked when something is perceived or interpreted as a danger, whether you confront it, run from it, or *worry* about it. But neither the body nor the mind effectively deals with the danger. There is usually neither escape nor physical victory over it. There is no peace. The worry or danger stays with us as bottled-up emotions such as anger or sadness and continues to drain energy. Recent research suggests that it drains more than energy. Our immune system is weakened by stress, too, and thus we are left more vulnerable to disease.

What can we do to plug the energy drain? Regularly expending in a constructive way the potential energy built up for fight or flight helps a great deal. Regular aerobic exercise performs this function. But can the reflex itself be changed, short-circuited before it prompts the body into the all-alert, stress-producing phase? Yes. The place to do this is in the mind at the very point where these events and activities are first perceived and interpreted as dangers. This is not a physical task, since these interpretations are not made on the physical plane. They are made on a plane that includes mental, emotional, and spiritual habits, beliefs, and activities. I usually refer to this area as metaphysics, because it extends beyond the physical. It may be called by many names, depending on whether you are a guru, minister, psychologist, or medical doctor.

No matter what you call this area, you must deal with

it before you can achieve a state of peace, health, and high energy. Just as our physical habits affect our energy levels, the way we take psychological, emotional, and spiritual care of ourselves can either energize us or bewilder and weaken us.

In the first part of this chapter, I will discuss the emotional and psychological habits and beliefs that tend to weaken us. Then I will discuss how to turn these habits around and learn new techniques to energize yourself from within.

ARE YOU A STRESS JUNKIE?

The stress junkie is a person who is addicted to stress and uses it to whip the adrenal glands, the same way many other people use coffee, sugar, alcohol, or whatever food they're sensitive to. Many people use stress in this way as well as abusive foods and chemicals. The stress junkie creates stressful situations, anger, and all kind of upset in his life, because when he's stressed, his adrenals pump out a little more adrenal hormones and give him a bit more short-term energy.

If you ignore this use of emotional stress or intentionally use it to pump up your energy, it will have the same effect as the use of artificial chemical stimulants. You will weaken your adrenal glands and make it much more difficult to regain or maintain true high energy. With good personal physical care, the problem may subside a bit temporarily, but unless your problem is strictly on the physical plane, you will set yourself up to re-experience it when any minor stress arises. Eventually you will experience the emotional overload commonly known as "burnout."

Burnout occurs when you reach a point where you have more going on in your head than you can handle. You create emotional dependencies and other reasons that seem to be believable enough to provide an excuse to be tired or sick. You short-circuit yourself and literally burn out all your energy. Either you suffer some debilitating disease that forces rest into your life, have a nervous breakdown, quit your job to go live in the woods, or find some other way to escape from the stress that's become too much for you.

ATTACHMENT DRAINS ENERGY

The main source of stress in most people's lives is attachment. You are attached if you insist that a given situation must be handled in a certain way and then become close minded to other possibilities, considering any other outcomes as failure. This rigidity will only lead you into more disappointments and upset. It will also tire you out.

It's not the problem itself that causes the stress, it's your reaction to dealing with the problem. In and of itself, a problem has no negative or positive meaning. It's just a situation that needs to be dealt with.

If you can deal calmly with the problem and do your best with the available resources, and then accept the outcome and learn any lessons you can about what happened, you will be able to minimize stress. If the problem should ever come up again, you'll be less likely to experience stress when dealing with it.

Let's define any emotional attachment or dependency as *sympathy*. I know that the normal definition of sympathy is somewhat different, but in this case, it does not mean thoughtful caring, it means potentially hurtful emotional attachment. Sympathy is the essence of what stress is—it means attaching yourself to an emotion, an idea, a person, place, or thing and allowing it to rule you, to determine what you think, feel, and do. You then either have a temporary excess of energy or plunge into fatigued depression as you get bounced back and forth, up and down, by people and situations that come along in your life.

If you react to life sympathetically, you are subjecting not only your mind but your body as well to considerable stress. Your adrenal glands have to put out extra hormones to deal with all the thrills and spills of the roller coaster ride. Some animal studies show that as the adrenals begin to tire and weaken in response to continued stress, they put out measurably diminished amounts of hormones. As we've seen previously, when the adrenals weaken, they take the whole body along with them. You'll then be more likely to suffer from blood sugar and food sensitivity reactions. You'll have trouble focusing and concentrating. Varying degrees of fatigue will take over your mind and body.

There is a chance that when the emotionally dependent person is made aware of himself, he will panic and become so "attached" to curing his problem that he will make a 180-degree emotional turn. Or maybe he has been "burned" in some situations in life to such a degree that he gets fearful of those attachments and overcompensates. He flees from attachment to detachment, a place where he doesn't care at all about any particular issue. This *apathy* is the flip side of sympathy. They're both attachments. Apathy is merely a strong attachment to not being attached. Both apathy and sympathy generate physiological stress, often in different forms.

Apathetic people may seem better off on the surface than sympathetic people because they don't seem to suffer. But it's doubtful they really are better off. Apathetic people tend to feel lonely and misunderstood in life. They tend to have few friends and to be very cynical.

What's more, because they tend not to ever show upset or pain even to themselves, it's hard to get them to realize there's anything to heal in the first place. Apathetic people are obvious to everyone but themselves. They have to swallow hard and admit, "Yes, this is me, and I'd better do something about it."

The position we're striving for is the center position between apathy and sympathy, the center of balance and peace. We can call this position *empathy*. As a situation comes along in your life, you will get a clear look at it, have a sense or a feeling of what it is about, and handle it as best you can without getting attached to the outcome or stewing over it for days. You will learn the lesson of the situation for future reference and move along with your life.

Naturally, no one is purely apathetic, sympathetic, or

empathetic. We're a combination of all three. In some situations we can be empathetic, no problem. In many situations, our response will be a three-part harmony, of which we hope empathy will be the dominant note.

BELIEF CREATES EXPERIENCE

What actually decides whether our response will be energy-draining or energy-building, sympathetic or empathetic, is our own belief about the experience. The meaning we assign to something determines how much energy we are going to use up dealing with it. If we assign an ominous, worrisome meaning to something, it's going to drain us. Internally, our beliefs create the emotional experience for us, which in turn creates the inner physiological response. If we assign the meaning "Danger, this is a threat!" to an event, the glandular stress response I described earlier will be stimulated and the event *will* actually become a danger. If we believe that performing a certain task is going to exhaust us, then sure enough, it will.

Externally, our beliefs can create the actual physical experience, too. Most people think the opposite is true, that experience creates belief; that as we go through life, we draw conclusions from our experiences and come up with beliefs. It is certainly true that we are constantly looking to deduce basic truths from our experience. But many of the beliefs that most strongly affect us and determine how we are going to live were formed as we were growing up, under the influence of parents, teachers, and other role models. Other beliefs were hastily thrown together in times of crisis to help make some sense out of a traumatic event. These usually bear little resemblance to truth, but they color our perceptions nevertheless.

Because you have a belief, you create through subconscious actions, or perceive through your physical senses, experiences that will prove you right for having that belief in the first place. You will self-fulfill your own most deeply held prophecies. Your beliefs will perceptually create the actual physical events in your life.

The idea that belief creates experience is not as "far out" as you may think. Research in physics has suggested that this concept is one of the central truths of the physical as well as the nonphysical universe. Physicists working with the most advanced and sophisticated equipment available have discovered that observation affects the outcome of any experiment, situation, or incident. It is impossible without "blinding," scientists now know, to observe something happening without affecting what is happening. "Blinding" is done in scientific experiments to prevent the observer and/or the subject from knowing which experimental group the subject is in, so neither's preconceived expectations can serve to influence the outcome of the experiment. The observer's expectations are a powerful influence on the outcome. If this is true in physics, which deals with basic concepts of the physical universe, imagine what the implications are for the metaphysical universe within our own lives!

For example, one widely held belief that creates an enormous amount of fatigue is: "Work is a drag. Vacations and time off are what I live for."

The many people who hold this belief hate Mondays and love Fridays. Or they take the apathetic point of view and try to ignore their dislike of work because it's something they have to do to make a living.

This attitude precludes the possibility of having a good time at work. Instead of learning and prospering at work, instead of doing an honorable and caring job, they merely occupy space. The lack of peace and happiness, the negative energy that results from this negative belief, becomes a drain on energy. This belief can ruin vacations, too. These people often have the expectation that what happens on vacation must make up for all the hell they've put themselves through on the job. A mix-up with hotel reservations or a day of rain will devastate them.

WE CREATE OUR OWN MEANING

Another fundamental belief that drains energy is the conviction that situations and things have inherent, absolute meaning. For example, "Waiting in line always means you're wasting time."

Nothing has absolute inherent meaning; we superimpose the meaning. We look at something and say, "This is good . . . or terrible . . . or wonderful." Some people can look at an experience and have a positive, "it's OK" understanding about the whole thing, while others suffer tremendously over the same situation. It is the eyes through which the experience is seen that are different.

A football game is an excellent example of this concept. Everyone in the stadium is watching the same game. But half will be upset at the end and the other half will be happy. Most people are rooting for their favorite team to the extent that they're attached to the outcome (i.e., whether their team wins) as well as the means (i. e., what happens during the course of the game). This is the sympathetic viewpoint. The fan has decided that his own life is somehow bound up with the fate of the team in such a way that the team's success is of great emotional importance. This person will experience temporary elation and depression with each turn of events in the game. This is frequently how many of us experience our entire lives—as alternating short periods of elation and depression, the roller coaster ride of stress.

A few people will assume an apathetic viewpoint. Perhaps they just went along for the ride, got free tickets, or couldn't care less about the outcome of the game.

The person looking through empathetic eyes might be looking forward to the game because he'll be out in the fresh air and he really appreciates watching skilled players perform. This person will enjoy the essence of the game, the spirit of competition as well as the excellence of the performance of all players. He may have a favorite team, but he will not become upset when "his" team makes a mistake or is bested by a superior performance by the other team. On balance, though certain events may thrill or excite him, he will view the entire game with a sense of appreciation, entertainment, and attention to interesting details. He will draw energy from his excitement, rather than let any attachments to "the way things have to be" drain his energy.

The point to remember is that *we make up all the meanings, so we can change our minds any time we like.*

WE HAVE A CHOICE

It is not necessary to go through life a slave to energy-draining attachments and beliefs. What is learned can be unlearned. Closed-mindedness can yield to open-mindedness. We can learn to see the infinite number of possible ways to turn unexpected problems into meaningful lessons. Here's an example that should strike a familiar chord in many of you:

You're driving to an appointment and you're barely on time. Your car breaks down. It's doubtful you can fix it, and you'd ruin your good clothes if you tried. You're going to miss your appointment. You feel helpless and stranded. You're also worried how much this will cost you before the car's back in running order. So you do what most people would do: you *choose* to be upset.

Most people would never look at it as a choice—"Do I want to get upset about this or do I not want to get upset?" They deem it an automatic upset. But there is always the choice of whether to be upset or not.

Let's compare two options and see which you like better:

Option 1. You choose to be upset. You rant and rave at the car and at the mechanic you finally contact, complain to the person with whom you had the appointment and his secretary, and vent what's left of your frustration on your family that night. Instead of everyone sharing the stress in a healing, forgiving manner, everyone becomes more tense. The negative effects of the original problem magnify and grow among everyone who comes in contact with you. All may seemingly forget the incident within hours or days, but the subtle and powerful memory of the upset will remain, especially if you have suppressed any anger or ill feelings inside yourself.

Option 2. You decide to be adventurous and choose not to be upset. Your first priority is to call the person with whom you had the appointment, explain, and reschedule. You will gain his respect and appreciation for keeping your composure.

Your car is towed to a service station. You go about getting the car fixed and even chat with the mechanic. You may have to call a friend or relative to pick you up, but you finally get home. You've dealt with the whole problem and you haven't gotten upset. Dinner is pleasant because rather than sharing an experience of anger, you can share an experience of survival and even triumph. Perhaps you share a story or two about the mechanic who fixed the car. There's peace in your life and also in the lives of the people with whom you come in contact. You've made a good impression on a few people and been enriched in the process.

By making a simple choice, as if by magic, you've turned pain into peace. You've truly made the best of the situation.

This example shows how attachment to a problem magnifies the problem. Being attached to the goal of getting to the appointment on time results in becoming upset when you can't make it. You're also attached to the means by which you would make it—having precise expectations that you would leave the house on time, that traffic would flow smoothly, and that the car would hold up.

In addition, you may have a whole range of other attachments—to the goal of having the actual meeting go just so, to your business dealings and personal relationships, to the raising of your children, the food you eat, the sleep you do not get, and every other aspect of your life. Your existence is saddled with an infinity of attachments that needlessly drain energy and take away from your enjoyment of life.

This self-manufactured torment is even worse than it appears, because stress doesn't only concern issues that we usually perceive as problems. You can be stressed by a positive circumstance as much as by a negative one. It's the nature of attachment itself that causes the stress, not the nature of what you're attached to. This is why a promotion, a wedding, or even Christmas are often just as stressful as a divorce or losing your job. Christmas, for example, becomes a stress when you become attached to your expectations of how the holiday *must* go. You hope for snow—but not enough to make driving hazardous. You worry about whether you bought the right gifts for everyone. You become attached to the fantasy that all your friends and relatives will really get into the Christmas spirit. In short, you desperately want the holiday to be as grand as your childhood memories.

Naturally, since childhood memories are largely fantasy, nothing ever quite lives up to such expectations. So you may experience a sense of loss. A holiday with a tremendous potential for happy times is tainted with anxiety, disappointment, and depression.

We can avoid all the worry, upset, disappointment, and depression surrounding events in our lives and find real peace within ourselves if we stop expecting external occurrences to create peace and happiness for us. This kind of thinking only sets us up for disappointment. What goes on in our lives will bring only bits and pieces of contentment that are short-lived, and will further encourage us in the mistaken belief that peace comes only when everything happens exactly according to plan.

If we instead have peace inside of us, as we look out at the world we can always make any situation peaceful and all right for us.

TAKE YOURSELF OFF AUTOMATIC PILOT

If you don't pay attention, you'll keep having the same upsets over the same problems all your life. What if every time you have to wait in line you automatically get upset? It's a relatively minor problem that happens every day—at the supermarket, in traffic, at the bank. Standing in line just happens to make you upset. Why? Because you believe that it does. As long as you look at it that way, you're a victim of the external circumstance and you'll never see a way out. You will automatically get upset whenever you have to wait in line.

There's a little program or prerecorded message in your brain that clicks in whenever you have to wait. It causes you to sweat and fuss and fume. It's based upon some belief. In order to reprogram your brain, you need to uncover the specific belief. In this case, perhaps it has something to do with fear of dying, fear of not getting everything done in your life that you want to do, so you can't afford to waste even a minute. Or perhaps it's a feeling of insecurity that makes you uneasy whenever you feel that your identity might be lost in a mass of human bodies.

Once you eliminate preprogrammed beliefs that predispose you to automatic upsets, you will find experiences that once caused you upset may now be no trouble at all and may even be a new source of life, peace, and happiness. You will not be a victim anymore. You will have a sense of your own true power. You will also be free of fatigue-causing beliefs.

Here's a common fatigue-producing belief: "If you don't have to struggle for it, it's not worthwhile." This usually works in tandem with another belief: "Working hard means I'm going to be exhausted at the end of the day." People with these beliefs experience a lot of struggle and use up a lot of energy, all in order to feel that they've done something worthwhile by the end of the day. They actually create a lot of fatigue for themselves.

Not all fatigue-causing beliefs are long-term. You might wake up one morning and fall victim to the temporary belief, "I'm really tired, I'll never be able to make my customary two-mile run this morning." With this belief, you probably won't.

You can just as easily say to yourself: "I'll start the run and just see what happens as I go along. If I have to, I'll just walk home and not be upset that I didn't get the entire two miles in."

Instead of creating beliefs that impose limits on yourself, create beliefs that allow you to do your very best. Allow your life's progress to be unlimited. You'll never fulfill your potential if you don't believe you can do any better. Don't talk yourself down with your beliefs, build yourself up. You'll be surprised what you can accomplish with an enthusiastic voice inside cheering you on. Before you can really get the most out of your own internal cheering section, however, you need to deal effectively with the voice inside you that wants to steal your energy.

BEWARE OF THE TROUBLEMAKER WITHIN YOU

Why are these negative beliefs so powerful in our lives, even when we know about them? What enables them to continually hound us? We all have a troublemaker within us: the negative ego. This part of the ego tries to keep our negative beliefs in the forefront of our awareness. I call it the negative ego to differentiate it from the useful ego, the part of the ego that retrieves information for our consideration. The negative ego does not serve, but seeks to rule.

Why do we listen to the negative ego? Because in our ignorance of the positive vitality of life, it often is the only thing that makes us feel alive, although the misery it causes sometimes makes us wish that we weren't. The negative ego leads us into powerful emotional attachments that can give us a real energy charge. But these intense passions and pains are only temporary, and ultimately they drain us of energy.

The negative ego also loves power, and it has learned that we get sucked in by its constant chatter, its "yakkity yak." It gives us interpretations like: "This person is attacking you! You're being victimized! You have to defend yourself! You deserve more!"

It is the constant paying attention to this "mind chatter" and trying to interpret it and sort it all out that are so exhausting and leave us so tired.

On the surface the negative ego may appear to be protecting us from the world. But it's not looking out for our best interests. It may deliver some short-term charge of energy—as a candy bar does—but it will always cause long-term loss, since loss and victimhood are the feelings from which it draws power. And if we continue to entertain these feelings and negative beliefs they will become self-fulfilling, self-perpetuating pains and upsets in our life.

When you recognize the voice of the negative ego whispering nonsense in your ear, simply tell it to shut up and go away. I use a little trick when I catch myself listening to my negative ego's voice. I say, "I'm not interested in your interpretation of this. Get out of here!" and tug on my ear. That ear tug is the off switch for the negative ego's message.

The negative ego is not going to give up and slink away whimpering. It has had a lot of power over your life the last 20, 30, 40, or more years. It's going to fight you, kicking and screaming, all the way. So you'll have to keep your guard up and be diligent for a while.

It is important, even when putting your negative ego in its proper place, to treat yourself lovingly. Don't attempt to heal yourself with a vengeance. If you're not doing it with empathetic eyes, you're not learning the lesson as well as you might, and you won't accomplish your goal as well as you can.

CREATIVE VISUALIZATION FOR ENERGY

Sometimes your negative beliefs are so deeply embedded in your consciousness that they defy attempts to uproot them. This can be effectively handled by a useful therapeutic tool called *creative visualization,* in which you go into a light meditative state and create a mental picture of what you want to happen.

When they first hear about this method, many people don't think it can be effective. It sounds too simple, they believe, to really deal with deeply rooted, emotionally charged beliefs. In truth, this is an extremely powerful tool. Bruce, a 58-year-old college professor, came to me complaining of fatigue, backache, and pains in his joints. His energy level was very low. Bruce had recently experienced a severe stress: an older child of his had died. Bruce was extremely distressed over the death, unable to live at peace with it. We

talked about the situation a lot, and I taught Bruce a visualization in which he could heal the relationship with his deceased
child.

Bruce needed to develop a sense that his life had to go
on and that he could choose between pain and peace. He
needed not only to heal and complete his relationship with
the child, but also to heal himself by removing his guilt for
having been unable to prevent the death. This was also to be
accomplished through visualization.

Was visualization a strong enough tool to deal effectively with such a profound sense of loss and guilt? It was.
Bruce's energy returned, and his aches and pains improved.

Here is a simplified visualization,[6] which can be used
to change beliefs and heal. Let's say you want to change the
belief that work makes you weary and that it's impossible to
have enough energy or life force to achieve a certain goal.
If you are going to change this belief, what will you replace
it with? Always create a positively constructed belief, not a
negative one. In this case, you would say, "I can have all the
energy I want and need to achieve my goals." You would
not say, "I will not limit my expectations of how much energy I can have," because that is a negatively constructed
statement.

Find a time at home when you won't be interrupted.
Turn off the phone and close the door. Sit up comfortably in
a straight chair with your eyes closed and begin to concentrate on your breathing. Notice the sound the air makes
going in and out of your nose, or pay attention to your chest
wall moving slowly up and down. Relax . . . relax . . . relax.

Focus on your breathing. As you feel yourself relaxed
yet focused, imagine yourself in a dark hallway. You're walking down the hallway. And even though it's pitch black, you
seem to have a sense of where to go. Intuition, not sight,
guides you down that long hallway.

All of a sudden you see a soft light at the end of the
corridor. Near that light you see a door. You go through the
door into a room. In the center of the room is a table. On the
table is a candle, the only light in the room.

You walk over to the table and notice a book in the candlelight, a beautiful old leather-bound book. It's a bit worn—you can tell it's been opened and closed a lot. But it's still lovely and has a good, warm quality. You pick up the book and hold it. This is your Book of Beliefs.

You turn to a page in the book. It's the page that contains the belief you want to change. You clearly see the belief written on the page. Interestingly enough, all the pages in this book are perforated on the binding side. And so all you have to do now is take the page, hold it firmly with your thumb against the perforations and gently and neatly tear the page out of the book.

You do so, and then look at the page. You see that this belief doesn't work for you. You say to yourself: "This belief does not work in my life anymore." You hold the page so that its edge touches the candle flame, and you let it catch fire. When the page is nearly consumed, you set it in a round crystal ashtray on the table and watch it burn to ashes. You now know that this belief has been absolutely eliminated from your Book of Beliefs.

Now you turn to a clean page in the book. On the table is a fountain pen of fine quality. You make up a new belief to replace the old one—a positive, peaceful, *constructive* belief that now works for you. You take the pen and write this new belief neatly on the blank page. You read the new belief and tell yourself: "This new belief will work for me now in my life and produce peace and happiness."

It doesn't mean that at some future time you won't have to change this belief, too. But if you're looking through clear, empathetic eyes, generally this belief will work for you most of your life.

The ink has dried. The new belief is indelibly imprinted in your Book of Beliefs. Close the book. Set it down on the table. Turn around, walk out of the room. Close the door behind you and start walking down the hallway. Again you know exactly where to go even though it's dark and you've turned away from the lighted area near the door. Take a few more steps. Now focus back to your breathing with your eyes still closed. Count backwards slowly from 5 to 1 and know that at the sound of 1 you will open your eyes and be relaxed and alert and feel fine.

Realistically, doing this visualization exercise does not mean that your old belief will never be a problem for you again. For a while, you may have to gently remind yourself of your new belief. Don't give yourself a hard time about not being perfect when the old belief crops up. You will begin to say to yourself when you're upset over a specific issue, "Hey, that's a belief that I've changed, so I don't have to be upset anymore. I can deal with this problem in peace." When you find yourself getting weary in a psychological or emotional way, you will be able to say to yourself that you've changed that belief and you do not have to be fatigued anymore.

Does this mean you will no longer have to deal with the physical causes of fatigue? Hardly. If anything, the physical causes will become more apparent to you as you realize that they are physically preventing you from having all the energy you can now feel from within. This will help motivate you to improve the way you take care of yourself.

You will start to catch yourself earlier and earlier when certain fatigue-causing issues surface in your life. Soon, you will be telling yourself, "I'm not even going to start to get upset or tired about this!"

You may have to give yourself a kind talking-to once in a while, or revisualize the new belief in your Book of Beliefs. You're literally retraining yourself into a new mode of thinking. But pretty soon you'll find yourself getting less and less upset or fatigued about that issue, less and less often. You'll soon find it fading from you more and more quickly, until eventually it's not happening at all. That's the point when you look back at something that happened half an hour ago and think, "Wow! Six months ago I would have gotten really upset about that!" Or, "I would never have believed that I could handle that problem so gracefully!"

MEDITATION ALSO QUIETS THE
TROUBLEMAKER WITHIN

Meditation is a formal exercise in turning off the negative ego's voice and quieting the critical, judgmental mind. That's why meditators enjoy their meditation so much: they get to spend time with the part of their being that's really their essence and their source of infinite peace and happiness.

To begin to meditate is to take power away from the negative ego. So when you sit down to meditate, the negative ego is going to be horrified. True to its seductive, deceptive nature, it will whisper in your ear, "What are you doing this for? This is a waste of time! Come on, we have better things to do!"

If you're not strong against the negative ego, you may finally accept the belief that meditation doesn't work for you, and you won't get anywhere with it. If you are diligent and patient with yourself, you *can* learn to meditate and profoundly increase the peace, happiness, and energy in your life.

The meditation I'm going to teach you is a basic yoga meditation, a breathing technique that's very easy to do. Always choose a quiet time and a quiet place where you won't be interrupted. Sit up either in a straight-back chair or against the wall on a cushion. Rest your hands comfortably in your lap, not touching each other. Don't cross your feet unless you're sitting on the floor. Be comfortable.

Close your eyes and start to pay attention to your breathing. You can focus on the breath going in and out through your nose and the sound it makes, or you can focus on the rising and falling of your chest wall or abdomen, or any other aspect of your breathing that you choose. Pick one and focus your total concentration on that aspect.

If you notice that the negative ego's voice has stolen your concentration away from your breathing, don't give yourself a hard time about it. Just give your ear a tug if you wish and come back to the breathing. The first couple of times you try it, you may get only 3 to 5 minutes done because the voice will be popping in there after every couple of breaths, distracting you frequently.

As you attempt to achieve quiet it becomes very obvious how constant the chatter of the mind is and how much time and energy you spend listening to it throughout your life. This realization should motivate you to further practice.

When you do become distracted, stop if you feel you can't effectively go on. Come back to meditation later in the day. Do the best you can, and stick with it. Some people have success quickly, some take more time. Be patient, and just keep yanking on that ear. Pretty soon you'll notice the voice pops up less often, you'll notice it sooner, and it will go away more quickly. One day you'll find you've been meditating for 20 minutes and it really didn't seem that hard.

I suggest you try initially to meditate 15 to 20 minutes twice daily. That's the best time you'll spend in the whole day. If done first thing in the morning, it will set the course for the day. It will create the clear, empathetic eyes that you want, which will keep you in a peaceful temperament throughout the day.

As you enjoy it more and appreciate how it works for you, you may want to meditate longer than 20 minutes. Down the line you will actually gain a sense of timelessness while meditating: you'll be sitting there thinking you've taken 10 or 15 minutes, when 45 minutes have gone by.

First thing in the morning is a great time to meditate, because you've just awakened and there hasn't been a lot going on for your negative ego to bother you with. You'll tend to be a little clearer in the morning and a little more likely to have success in your meditation. So get up, have a Vitamin C Drink, shake the cobwebs out, and then sit down and try out the meditation.

Bedtime is another good time to meditate. It will sweep away all the problems of the day and relax you so that you're ready to sleep well.

Try meditating when you find yourself getting upset. It may be harder to meditate at this time, but it's well worth the effort. It's one of the best ways to avoid an emotional melodrama that you will regret later.

As you learn to meditate, always focus on the positive qualities of your goal. This will give you confidence and automatically encourage you. What I've just taught you is a technique. Like any fine musical instrument or sport, it must be practiced diligently to be done successfully.

The feeling you'll have when you finish meditating will be quite a reinforcement in itself. You will have a sense of well-being. You'll feel energized. In a certain sense, you will feel "high."

Many people drink or take drugs in order to reach a similar state of consciousness. The alcohol or the drug is often used as the vehicle that takes them out of the ego self and gives them glimpses of the spiritual self, the part of the self that does not fall victim to hurtful emotional dependencies and attachments. They get to see how it feels to look through empathetic eyes. They feel relaxed and happy. Things don't upset them. They have energy. They feel creative.

Frequently, because of the abusive nature of many of the substances they use, they go beyond or around the peaceful state they seek. Not only that, but they also get attached to the vehicle, the drug itself, which becomes the only way they know to reach that state of mind. A peaceful, calm, creative state of mind can be achieved without alcohol or drugs. Use meditation and creative visualization to create a "high" from the inside out.

Are you wondering what you'll become as you work toward your goal? Unlike the negative ego, which sees the world as a series of attack-defend strategies in a lifelong war against the world, the spiritual self, or the soul, creates situations where everyone who participates wins. There is no loser, there is no battle. The soul can afford not to be aggressive or defensive. It can be vulnerable because it knows it really cannot be hurt.

If you ever wonder which voice you're listening to and being directed by, it's easy to tell. The voice of the negative ego creates pain, upset, suffering, and a sense of loss. It is judgmental and believes that there isn't enough time. It sees you as better or worse than other people. The voice of the spiritual self, on the other hand, sees peace, calm, joy, love, happiness, and forgiveness. It knows there's plenty of time and sees you as equal to other people. Ask yourself which of these sets of qualities you are experiencing and you can tell which voice you're listening to. If it's the ego's voice, you can make a conscious choice to quiet the mind and let the soul's voice be heard.

HAVE FUN ALONG THE WAY

Fun, humor, and situations that are lighthearted and create laughter are very healing for the soul and the adrenal glands. The delight we feel when we have fun is an example of the kind of uplifting spirit and energy we have hidden away inside us. Most of the time we take our lives too seriously. There's also a great tendency when we're ill or tired to take everything too seriously. It is crucial to maintain the attitude of fun, to remember to have a good time, to be lighthearted and happy as you heal yourself and make progress in rebuilding your energy. You'll heal more quickly.

One of the reasons we lose energy as we get older is that we lose the child's natural ability to have fun and to spontaneously create fun. I would urge everyone to become more lighthearted, to have more fun and take everything much less seriously. It's not as hard as you may think. You'll get just as much done, you'll be just as efficient, and you'll have a lot more peace and happiness. Your body will work a lot better and you'll have more energy.

Such was the case for Louis, a 40-year-old man who complained of an overwhelming amount of stress in his life. He had financial worries and his business had not been doing well. Worst of all, his relationship with his teenage son was not satisfactory and was causing a lot of strain. Father and son often disagreed and yelled at each other. Louis was very tired from all of this.

I told Louis he ought to just go out and have a good time with his son. I told him that it wasn't necessary for them to see eye to eye on everything and that it was perfectly reasonable to disagree. But it was definitely not necessary for them to disapprove of each other, to take love away from each other. So I advised that they go out once or twice a week and play baseball, or go to a movie, or whatever—just to have fun with one another. I also recommended that Louis take a yoga class and learn to quiet his mind with meditation and yoga.

When Louis took my advice, he and his son improved their relationship quite a bit. It wasn't yet perfect, but they had a better sense of understanding between them. They could live with each other and work things out. Louis had a lot more peace in his life—and a lot more energy.

Steps to Energizing Yourself from Within

Step One: Make a list of your energy-draining beliefs. Start with the most obvious ones, some of which were mentioned in this chapter. You may be able to readily list a dozen beliefs that limit and weaken you. Or it may take you some time to examine your mind and become aware of the beliefs under which you operate. Write down any belief that limits or deprives you in any way. Some of these beliefs may seem reasonable to you. Leave those alone for now. What we want to do is go after the beliefs that are obviously weakening ones. Using the process described in this chapter, change those beliefs to their positive, liberating counterparts. Begin with the one that is the most limiting, then proceed through the list at a pace that is comfortable and productive.

Step Two: As you notice your new beliefs changing your behavior and increasing your energy, you will be motivated to go on and change more beliefs. Move down your list. You may find that your feelings about some of your beliefs will change. Beliefs that may have seemed reasonable to you a few weeks ago may no longer seem worth keeping. Let them go and replace them with energizing beliefs.

You may also begin to become aware of weakening and limiting beliefs that you were not aware of before. Add these to the list and replace them in turn.

Step Three: If you have a belief that is preventing you from meditating to increase your energy and peace, you can now replace that belief with one that will allow you to bring this resource into your life.

Likewise, if you have beliefs that have prevented you from making use of any part of this book to create high energy for yourself, you can now begin to replace those beliefs and work your way toward true high energy.

HIGH-ENERGY TIP

FOCUS

The ability to focus your complete attention on what you're doing while you're doing it is what ultimately brings peace. Most of us, most of the time, are busy doing one activity but have our mind in five other places at the same time. This separation of action and thought creates tension and drains us of energy.

Often the reason people are drawn to certain sports, hobbies, crafts, and performing arts is that when they are engaged in those activities, they are naturally able to totally focus their physical, mental, and spiritual energy in the moment. As your feelings of success, fulfillment, joy, and happiness increase, you will naturally associate these feelings with the activity, which becomes important to you. In truth, it's not the activity itself that necessarily creates all these fine feelings—it's the ability to focus. The more the ability to focus can be brought into all aspects and activities of your life, the more peace and energy you will enjoy.

This focus will generally have a timeless quality about it.

That is, while engaged in the activity that helps you focus, you will frequently lose track of time. This sense of timelessness is an indication that you are achieving focus.

The ability to focus your mind, body, and spirit in the moment—no matter what you're doing—creates a profound sense of peace, increases your ability to use the natural energy around you, and leaves you feeling energized.

10

Fine-Tuning for Your High-Energy Future

"Life does not teach you.
It provides you the opportunity to learn."

LAZARIS

Healing is a process that goes on throughout your life. It's a progressive fine-tuning and paying attention to details. By observing what's going on and how it affects you, you can discover what specific things you do in your life that help you feel good and increase your energy and give you more peace and happiness—and what things don't. By doing that fine-tuning as you go along, you create energy more and more consistently.

It's a process that doesn't necessarily have a lot of drama to it, but it will build confidence in your own ability to take care of yourself. I tell people all the time: Good health is not a gift. It's the result of good care. And you have to learn how to give yourself that care.

Keep the rules during the majority of your life. When you make an exception, you will be able to clearly see how deviating from that rule affects you. I'm not implying in this book that you should never watch TV, or try a new food, or taste a favorite old food once in a while. But you must pay close attention to see how you are affected by these activities. In a confused and fatigued body, this inner communication will be muddled. In your renewed, energetic, sensitivity-free body, the answers will be clear if you pay attention.

You will learn which exceptions don't cause trouble, so that you can indulge in them occasionally (once or twice a month), and which ones you really pay for by feeling fatigued or by other obvious symptoms that develop within 24 hours of indulging.

There is no right or wrong. *The whole point is to see what works best for you in your life.* The concept of discipline disappears, since your progress doesn't involve forcing yourself to make certain changes. As you recognize what things don't work for you, you will automatically begin to stop those habits. As you learn what gives you energy and happiness, you will develop and refine those habits. By paying attention over time, you will minimize habits that deplete and injure you and maximize habits that create energy and well-being.

It may be helpful to have some credentialed person guide you as you go along. But the best doctor and all the professional help that you can seek is nothing more than good coaching. And even the best of coaches can't run the race for you or live your life for you. Don't keep looking for someone else to pull some magic healing rabbit out of a hat. You created this fatigue. You had plenty of energy at one time. Fatigue didn't come from outer space and "get you" because you were standing in the wrong spot. Accept the responsibility for having created your present state of health; patient, heal thyself.

One of the reasons people get sick and one of the reasons they lose energy is that they give away too much energy, in the form of power over their lives, to too many different people and situations. They give it away to the government, to their boss, to their spouse or kids, they give it away to different people they deal with, like lawyers, accountants, and especially doctors.

You be in charge. Use all these professional people to provide you with information, all the pros and cons of the situation, then make a conscious decision yourself. If you make a poor decision, then learn from your mistake. Look for and understand the lessons to be learned, then move forward in your life as a wiser person.

The essence of accepting the responsibility lies not in blame but in the issue of power. In other words, if you're powerful enough to have gotten yourself into an illness or fatigue state, then you're just as powerful to re-energize yourself. You can heal the same condition that you created.

The way you get and hold onto health for yourself is by learning how your body works, learning how to take care of yourself, and building up your confidence in your own ability to heal and maintain health. And then you'll know you can do anything, not just have energy and feel good. You can heal anything. More important, you'll have absolute confidence as you look toward the future and realize, "I know how to be well, I know how to take care of myself now, I will not get sick." Believing that plays a major role in staying healthy. Belief creates experience.

DEVELOP A SENSE OF YOUR SELF

We are first and foremost here for our soul to go forward and learn the lessons it needs to learn. As we move toward this goal, we must make several choices along the way. We choose our profession, whether to marry and whether to have children, and we choose other responsibilities we wish to participate in. As we put our life energy into personally useful endeavors and into things that are meaningful to us—if we feel good about our life and what we're doing and we have a sense of reward, a sense of focus and creativity, and we do it with honor and integrity—we will be successful in our personal quest for fulfillment.

What can happen, however, if we're not careful, is this: along the way we get so caught up in our responsibilities that we forget about life's lessons. In a practical way we simply don't save enough energy to spend on ourselves, or the time that we truly need to pay attention to our feelings.

The classic example of this is seen in young, married, working mothers. They have many responsibilities: to be a wife, to spend time with and give attention to their husband, to cook and take care of him, and to look good for him. Frequently, they take on the additional time-consuming responsibilities that go along with raising children. They may also have a full- or part-time job. In addition, they may have community responsibilities, church or social groups. Since traditional roles have been restructured in recent times, many men find themselves in similar circumstances, or with the same amount of responsibilities in different areas.

When people have all of their energies going out in many different directions and there isn't enough left for themselves, the self frequently lets them know it. One way it lets them know it is by being fatigued, depressed, having a sense of frustration and of being a martyr—"I'm giving my life up for everyone around me and I have little energy that I can use for myself."

I'm not implying that you have to go through life being selfish. But all these responsibilities need to be balanced in life. You need to have enough energy left to use for those issues and desires that are personally fulfilling and that let you express your natural creativity. If you are a busy person who wears many hats, it is imperative that you create for yourself during the week personal free time with which you can do whatever you choose. Husbands and wives perhaps can take turns baby-sitting on Saturday or Sunday afternoon so that each can have some free time and energy.

As part of your sense of self, you must develop and express your creativity; otherwise a definite sense of frustration, accompanied by fatigue, will result. If all energies and time are "given up" to work for others, the inner self in a sense has nothing to be personally enthusiastic about for its own growth, activity, and imagination. Try to be creative at your work, whatever it is. It will make you more productive and more fulfilled in this part of your life. If you don't have the opportunity to use creativity at work, then make an effort to use it in your nonwork life.

SEEK OUT "HOLY COMPANY"

As you attempt to clean up your diet and break all your poor habits and create new healing habits, find "holy company." That is, associate with people who have that same mentality, people who will support your efforts to rebalance yourself, heal yourself, and have more happiness and energy. Don't hang out with people who have the old habits you're trying to break. If you do, you'll make it that much tougher to break those habits.

For example, if you're trying to break a smoking habit, don't spend time with smokers, but rather with nonsmoking friends. If you're trying to incorporate running, cycling, or some other aerobic exercise into your life, enter organized races and fun runs. Remember, a real friend will always support you in your efforts to be the best person you can be.

You must also be on the lookout for people with low energy and people with false "enervating energy." Whether they do it purposely or not, many people with low energy can be a drain on the people around them. If you must be around such people, it's important to maintain control over yourself so that you don't allow your energy to leak into situations created by the low-energy person. The classic example of this type person is the "friend" who frequently calls you up to dump all his or her troubles in your lap and complain about being a victim of life or a martyr, but who rarely if ever calls you with any positive energy. Those types of relationships are exhausting and need to be restructured or ended.

In the same way, people with nervous energy are more often than not actually low-energy people who, like hyperactive children, must hype themselves up to get through the day. Such people will also be an energy drain on you, since they inspire the same kind of nervous expense of energy. You may feel stimulated at first, but it will be the same kind of false stimulation that a drug or an allergenic food produces. Ultimately, you will use up energy just protecting yourself against such people. If you cannot avoid them, my advice is to keep them and their demands at arm's length, in the same manner that you keep energy-draining attitudes and beliefs from gaining control over you.

A person who has true high energy will not drain your energy. He or she will not make unreasonable demands on you and will support you rather than drain you. As you become more and more sensitive to the habits and attitudes that create energy for you, you will also become more aware of the people who create energy for you and inspire you to create it for yourself.

EBB AND FLOW

We know that many physical phenomena in the natural world travel in waves. Sound and light energy travel in waves. Our lives do, too. We have certain periods of progress where things are going very fast and well. We have places where we seem to plateau, places where things seem to slow down, and places where we ride the crest of the wave. There are also downside times, the downward side of the wave, where we seem to falter or lose a little energy or to have things not quite go our way—only to go into a trough where things steady out, followed by another upswing.

You're not going to feel better and have more wonderful things happen to you every day that you live. There will be upswing periods, and there will be downslide periods. Don't become emotionally dependent upon or attached to any particular direction of your energy ebb and flow. Don't throw out your new program or commitments or admit defeat because you're having an off day. Every active person who's living a busy life, who has responsibilities, and who generally has ample energy is going to have certain days when he has more or less energy than other days.

There will be days when the body just says, "Hey, I want to rest today." As you learn to take better care of yourself, you'll be able to trust these inner communications. On days when the body tells you to take it easy, for example, you might not want to exercise even though it's your scheduled exercise day. You just have to recognize and accept that.

Do what you can. Accept the progress that you make. Appreciate what is happening currently in your life and the lessons you are learning. Don't make today's achievement tomorrow's obligation. You might have a really great day where you get a lot accomplished. Don't obligate yourself to achieve more and more every day. Just stay in the moment, put forth your best and most conscious effort, peacefully accept whatever happens, and learn from it all.

HEALING IS A PROCESS

The spiritual teacher Lazaris has articulated the following guiding principle in regard to achieving goals: "The steps of getting there are the same as the qualities of being there."

A goal is not a point in time that you suddenly reach when everything you want happens. It's more dynamic than that. A goal actually represents a particular set of qualities that you want to achieve. And the qualities of being at the goal are the same qualities that are necessary to get to the goal. If you have a sense of the goal being a path and not a place or a time, then you can have a sense of achievement and ever-increasing confidence as you walk the path.

If you walk the path as conscientiously as you can, you will be a "realizing person"—a person who is on the way to his or her goals. It is not necessary to be a totally self-realized person before you experience peace and happiness in your life. This is perfection, and few have attained it. Being a self-realizing person is what we're after here. You will stay on your path as long as you continue to adopt for yourself the qualities of your goal. It is not an instant process but a life-long one. Every step you take with your goal of high energy clearly and honestly in mind is a step toward true high energy. Every step will teach you something, every step will in its own way help you create high energy.

BALANCE

As you attempt to initiate and refine all the suggestions in this book, you will need to pay attention in order to fine-tune the details. There will be many times when you need to take a close look at certain details. And there are also times when it is a good idea to step back and take a look at the whole picture of your life. Are all the details fitting and flowing together harmoniously to create a balanced picture of health and happiness?

If a little rearranging or restructuring of priorities is necessary to create harmony and balance in the large picture, you will be able to notice it from this perspective and make the appropriate adjustments. By always striving for an intuitive sense of balance in your life as you progress, you will create and maintain high energy, optimal health, and peace.

PERSONALIZED PROGRAMS

If you would like information about personalized programs or how to obtain high-quality, hypoallergenic nutritional supplements, you may address your correspondence to:

Rob Krakovitz, M.D.
c/o Jeremy P. Tarcher, Inc.
9110 Sunset Boulevard
Los Angeles, CA 90069

S U G G E S T E D   R E A D I N G

Here are some very good books that will provide you with additional information about some of the areas covered in *High Energy*. As effective as my program may be for you, I hope you will increase your knowledge and understanding of the ways of health.

General

Dr. Wright's Book of Nutritional Therapy, Jonathan Wright, M.D., Rodale Press, Emmaus, Pa., 1979.

The Main Ingredients, Susan Smith Jones, Biworld Publishers, New York, 1980.

A New Breed of Doctor, Alan Nittler, M.D., Pyramid House Publishers, New York, 1972.

Nutraerobics, Jeffrey Bland, Ph.D., Harper & Row Publishers, New York, 1983.

Nutrition Against Disease, Roger Williams, Ph.D., Bantam Books, New York, 1973.

Orthomolecular Nutrition, Abram Hoffer, Ph.D., and Morton Walker, D.P.M., Keats Publishing, New Canaan, Conn., 1978.

Your Health Under Siege: Using Nutrition to Fight Back, Jeffrey Bland, Ph.D., The Stephen Greene Press, New York, 1982.

Food Sensitivities and Detoxification/Desensitization

An Alternative Approach to Allergies, Theron Randolph, M.D., and Ralph W. Moss, Ph.D., Bantam, New York, 1980.

Brain Allergies, W.H. Philpott, M.D., and Dwight Kalita, Keats Publishing, New Canaan, Conn., 1980.

Dr. Mandell's 5-Day Allergy Relief System, Marshall Mandell, M.D., and Lynn Waller Scanlon, Thomas Y. Crowell Publishers, New York, 1985.

Fasting: The Ultimate Diet, Allan Cott, M.D., Bantam Books, New York, 1977.

Food Is Your Best Medicine, Henry Bieler, M.D., Vintage Books, New York, 1973.

Hypoglycemia: A Better Approach, Paavo Airola, Ph.D., Health Plus Publishers, New York, 1977.

Nontoxic and Natural, Debra Lynn Dadd, J.P. Tarcher, Los Angeles, 1984.

The Pulse Test—Easy Allergy Detection, Arthur F. Coco, M.D., Arco Publishing, New York, 1968.

Sugar Blues, William Dufty, Warner Books, New York, 1976.

Why Your Child Is Hyperactive, Ben Feingold, M.D., Random House, New York, 1974.

The Yeast Connection, William G. Crook, M.D., Professional Books, New York, 1984.

General Nutrition

The Albrecht Papers, William A. Albrecht, Ph.D., Acres USA Publishers, New York, 1982.

Food for Naught—The Decline in Nutrition, Ross Hume Hall, Harper & Row Publishers, Hagerstown, Md., 1974.

Recipe Books

All recipe books are suggested to provide menu ideas that are tasty and work to ensure your health. Be sure to leave out or find a substitute for any food or ingredient you are allergic to or even suspect you may be allergic to. Also, remember to use good judgment and common sense whenever preparing food. Leave out or substitute for any ingredient you know just isn't good for you.

Dr. Mandell's Allergy-Free Cookbook, Fran Gare Mandell, M.S., Pocketbooks, New York, 1984.

The Egg-Free, Milk-Free, Wheat-Free Cookbook, Becky Hamsick and S.L. Weisenfeld, M.D., Harper & Row Publishers, New York, 1982.

The Good Grains, Editors of Rodale Books, Rodale Press, Emmaus, Pa., 1982.

Laurel's Kitchen, Laurel Robertson, Carol Flinders, and Bronwen Godfrey, Bantam Books, New York, 1978.

Moosewood Cookbook, Mollie Katzen, Ten Speed Press, Berkeley, Calif., 1977.

Recipes for a Small Planet, Ellen Buchman Ewald, Ballantine Books, New York, 1975.

Tassajara Cooking, Edward Espe Brown, Shambhala, Westminster, Md., 1973.

The 20-Minute Natural Foods Cookbook, Sharon Claessens, Rodale Press, Emmaus, Pa., 1981.

Supplements

Biochemical Individuality, Roger Williams, Ph.D., Texas Press, Austin, 1969.

Earl Mindell's Vitamin Bible, Earl Mindell, Rawson Wade Publishers, New York, 1980.

The Healing Factor, Irwin Stone, Ph.D., Grosset & Dunlap, New York, 1972.

Nutrition and Vitamin Therapy, Michael Lesser, M.D., Bantam Books, New York, 1980.

The People's Guide to Vitamins and Minerals from A to Zinc, Dominick Bosco, Contemporary Books, Chicago, 1980.

Supernutrition, Richard Passwater, Ph.D., Pocketbooks, New York, 1978.

Vitamin C—The Common Cold and the Flu, Linus Pauling, Ph.D., Berkley Books, New York, 1982.

Exercise

Aerobics, Kenneth H. Cooper, M.D., M.P.H., Bantam Books, New York, 1968.

Bikram's Beginning Yoga Class, Bikram Choudhury, J.P. Tarcher, Los Angeles, 1978.

The Complete Illustrated Book of Yoga, Swami Vishnudevananda, WSP, Akron, Ohio, 1972.

Fit or Fat?, Covert Bailey, Houghton Mifflin Co., New York, 1979.

The New Aerobics, Kenneth H. Cooper, M.D., M.P.H., Bantam Books, New York, 1970.

Stretching, Bob Anderson, Runners World Books, Mountain View, Calif., 1980.

Energizing Yourself from Within

Anatomy of an Illness, Norman Cousins, W.W. Norton & Co., New York, 1981.

A Course in Miracles, The Foundation for Inner Peace, Tiburon, Calif., 1975.

Getting Well Again, O. Carl Simonton, M.D., and Stephanie Matthews-Simonton, Bantam Books, New York, 1978.

Handbook to Higher Consciousness, Ken Keyes, Jr., Living Love Center, Topanga Canyon, Calif., 1980.

The Healing Heart, Norman Cousins, Avon Books, New York, 1984.

Love Is Letting Go of Fear, Gerald Jampolsky, M.D., Celestial Arts, Berkeley, Calif., 1979.

The Relaxation and Stress Reduction Workbook, Martha Davis, Matthew McKay, and Elizabeth Eschelman, New Harbinger Publications, Oakland, Calif., 1982.

The Road Less Traveled, M. Scott Peck, Touchstone Books, Simon & Schuster, New York, 1980.

Stress Without Distress, Hans Selye, M.D., New American Library, New York, 1975.

Your Erroneous Zones, Wayne Dyer, Ph.D., Avon Books, New York, 1977.

REFERENCES

1. D.C. Costill, *et al.*, "Effects of Caffeine Ingestion on Metabolism and Exercise Performance," *Medicine and Science in Sports,* 1978, 10(3), pp. 155–158; and C.J. Estler, H.P.T. Ammon, and C. Herzog, "Swimming Capacity of Mice After Prolonged Treatment with Psychostimulants; 1. Effects of Caffeine on Swimming Performance and Cold Stress," *Psychopharmacology,* Springer-Verlag, Vol. 58, 1978, pp. 161–166.

2. *Nutrition Reviews,* Vol. 37, Feb. 1979, p. 38

3. *Vitamins and Hormones,* Vol. II, 1953, p. 133

4. *Current Therapeutic Research,* Vol. 4, March 1962, p. 98; "Treatment of Fatigue in a Surgical Practice," *The Journal of*

Abdominal Surgery, May 1962, p. 76; P.E. Formica, "The Housewife Syndrome: Treatment with Potassium and Magnesium Salts of Aspartic Acid," *The Journal of Abdominal Surgery*, April 1962, p. 73; and "Effects of Potassium-Magnesium Aspartate on Endurance Work in Man," *Indian Journal of Experimental Biology*, Vol. II, Sept. 1973, pp. 392–394.

5. Hans Selye, *Stress Without Distress*, New American Library, p. 31

6. Based upon the Book of Beliefs Visualization taught by Lazaris.

INDEX

ABOUT THE AUTHOR

Rob Krakovitz, M.D., was raised in Abington, Pennsylvania. He is a Phi Beta Kappa graduate of Pennsylvania State University and was graduated from George Washington University Medical School in Washington, D.C., in 1974.

After his internship, Dr. Krakovitz studied the methods of Henry Bieler, M.D., one of the early masters of detoxification theory. Following that he worked directly with Alan Nittler, M.D., of Santa Cruz, California, and Joseph Walters, M.D., of Sherman Oaks, California—two of the leading pioneer physicians in nutritional medicine.

Since 1977, Dr. Krakovitz has had his own private practice in Los Angeles, specializing in metabolic nutrition and wholistic medicine. His other specialty interests include food sensitivities, substance abuse recovery, and maximizing performance for athletes.

He is currently on the board of directors of both the International College of Applied Nutrition and the Orthomolecular Medical Society.